BOSTON
PUBLIC
LIBRARY

CHEMICAL FIRE APPARATUS

The records of the National Board of Fire Underwriters show that 80 per cent. of our fires are extinguished by the use of chemicals.

The remarkable success of chemical apparatus has led to an extensive use of this means of fire protection. Garages, stores, factories, industrial plants, private estates, rural communities—all find the chemical engine an indispensable safeguard.

The merits of chemical fire apparatus may be summed up briefly: This apparatus takes up little space and is always ready for immediate service; it is light, easily moved, and gets into service quickly; it requires but one or two men to handle; it is so simple any man of ordinary intelligence can understand it; it extinguishes fires with dispatch; it is particularly efficient on oil fires where plain water is almost useless; it saves water damage, which is often greater than the actual loss from fire.

The cost of maintenance of chemical apparatus is very small, as practically the only item of expense is the chemicals, which are commercial commodities easily obtained.

For these reasons a chemical engine in its various modifications has become an indispensable factor in every properly equipped fire department, and an absolute necessity in small towns and villages where the water facilities are limited.

An up-to-date modern factory is not complete unless one or more of these chemical engines is part of the equipment.

We guarantee advantages to purchasers that cannot be obtained elsewhere. We agree that all material and workmanship shall be of the best character obtainable, and will, at our own expense, replace such parts as may fail, where failure is attributable to defective material or inferior workmanship; and we agree that the apparatus and equipment will perform efficient duty, accident excepted, when properly and fairly handled.

CHEMICAL FIRE ENGINES

For half a century chemical fire engines saved millions of dollars worth of property in spite of the fact that they never did perform as advertised.

W. FRED CONWAY

FIRE BUFF HOUSE
DIVISION OF
HOME SAFETY EQUIPMENT CO., INC.
NEW ALBANY, INDIANA 47150-0709

Published by Fire Buff House
Division of
Home Safety Equipment Co., Inc.
New Albany, Indiana 47150-0709

Copyright © 1987 by W. Fred Conway

All rights reserved. No part of this book may be
reproduced in any form or by any electronic or
mechanical means including information storage
and retrieval systems without permission in
writing from the publisher, except by a
reviewer who may quote brief passages in a review.

Manufactured in the United States of America

"It was Shakespeare who said that
> 'A little fire is quickly trodden out,
> Which being suffered, rivers cannot quench.'

I have yet to learn that Shakespeare contemplated the introduction of chemical engines at the time he wrote those lines.

My experience with chemicals was of several years duration, and during the interval I attended many fires, over sixty percent of which were extinguished by chemical engines without assistance from the other portion of the department called."

Captain E. F. Martin, Boston Fire Department, 1886

Note: The Boston Fire Department was operating ten chemical engines in 1886, and ordered two more in 1890.

CONTENTS

Introduction 9

Chapter I   Chemicals Enter The Firehouse 11

Chapter II   The Hype That Sold Chemical Engines 28

Chapter III The Amazing Acceptance of Chemical Engines 32

Chapter IV   The All-Chemical Engine Fire Department 43

Chapter V   Hand Drawn Chemical Fire Engines 47

Chapter VI   Horse Drawn Chemical Fire Engines 69

Chapter VII Motorized Chemical Fire Engines 87

Chapter VIII The Most Unusual Chemical Fire Engine 121

Chapter IX   Chemicals Leave The Firehouse 125

Chemical engines, whether hand drawn, horse drawn, or motorized ranged from "plain Janes" to ornate rigs complete with carriage lamp, as this 1881 Holloway horse drawn double tank engine.

INTRODUCTION

Just what *is* a "chemical fire engine?" The name implies a fire engine which uses chemicals to extinguish fires, but this implication is somewhat misleading on two counts.

First we need to define "fire engine" and even more basically, "engine." Webster describes an engine as "any mechanism or machine designed to convert energy into mechanical work." The old hand operated fire engines used muscle power to pump water; steam fire engines of course used steam pressure to pump the water, and motorized fire engines used internal combustion engines to pump the water.

What energy source, then, pumped the water or "chemicals" in more than ten thousand chemical engines which were in use from 1872 through the 1920's and even into the 1930's? It was carbon dioxide under pressure of as much as 200 pounds per square inch, produced by a simple chemical reaction that every high school chemistry student soon learns:

$$H_2SO_4 + NaHCO_3 \longrightarrow CO_2$$

Sulphuric acid, often called "oil of vitriol," when combined with bicarbonate of soda (ordinary baking soda) forms an immediate chemical reaction producing what yesterday's firemen called carbonic acid gas, which we know today as carbon dioxide.

The pressure of this gas within the cylinder or chemical tank of the engine forced the soda water or "chemicals" onto the fire. Thus a fire department chemical engine was an engine which used not manpower nor horsepower nor steam nor internal combustion, but the *energy of a simple chemical reaction . . . . a true chemical engine!*

Secondly, what about the implication that the "chemicals" put out the fire? This was probably the biggest hoax or myth, though probably unintentional, that has ever deluded the fire service. How such a misnomer could have endured for half a century is almost beyond belief; for although the chemical engines were widely advertised and believed to extinguish fires up to 30 or 40 times more effectively than water, they were, in fact, *virtually no more effective on fire than plain water!*

But make no mistake . . . fire department chemical engines were of inestimable value. They were fast, were already charged and ready to activate within seconds after arriving at the fire, were dependable, and enabled firemen to get water on a blaze from three to ten minutes sooner than hand engines or steam engines. For decades they were the fast attack mini pumpers of the fire service.

A survey conducted in 1895 by fire insurance interests in Louisville, Ky. (reproduced in Chapter III) indicated that chemical engines were overwhelmingly accepted by fire departments everywhere, who used them to extinguish from 50% to 90% of *all* fires, and in a few cases 100%, where they were obviously the only type of fire engines available. The unusual story of the Baltimore County (Maryland) Fire Department's exclusive use of chemical engines is recounted in Chapter IV.

For a number of years manufacturers could scarcely keep up with the demand for chemical engines, which were standard equipment not only in the smallest towns and villages, but in cities of all sizes up to and including New York City, Chicago, and Boston, which pioneered their use.

From their entrance into the fire service in 1872 with hand drawn and horse drawn apparatus, on into the motorized era of the 1910's and 1920's (even into the 1930's) until their demise by the booster system and the final realization that "chemicals" were not better than water, the fire department chemical engines wrote for themselves a major and significant chapter of fire service history.

Although chemical engines have been largely upstaged over the years by the more glamorous hand and steam fire engines, hopefully this book will help the gleaming copper and brass chemical engines of yesteryear to regain their rightful place in the romantic and adventurous history of the American fire service.

I

CHEMICALS ENTER THE FIREHOUSE

$$H_2SO_4 + NaHCO_3 \longrightarrow CO_2$$

Although the first chemical fire engines were built in Chicago and Baltimore by Babcock and Holloway respectively, the idea originated in France.

In 1864, for the first time, Dr. F. Carlier and A. Vignon, an engineer in Paris, used sodium bicarbonate dissolved in water, to which acid was added. The resultant chemical reaction, which generated copious amounts of carbonic acid gas (carbon dioxide), soon led to the "soda-acid principle" of fire extinguishment.

Just four years later in 1868 the New York City Fire Department had soda-acid fire extinguishers installed in every fire station. These extinguishers were placed next to the watch desk, and were carried by firemen who ran with them to minor fires.

The true forerunner of the chemical engines, which were to prevail as an important arm of the fire service for more than half a century, was a "fire extinguisher wagon," carrying ten hand-operated soda-acid extinguishers and placed in service in Boston in 1871. It was designated "Extinguisher Co. No. 1." Evidently No. 1 met with considerable success, since during the next year No. 2 and No. 3 were placed in service, each carrying no less than twenty-five soda-acid extinguishers.

But so far as is known, Boston's three extinguisher wagons were the only ones ever placed in service because that same year, 1872, the very first chemical engine (really a huge soda-acid fire extinguisher on wheels) was manufactured by the Babcock Manufacturing Company of Chicago and placed in service in the Chicago Fire Department.

Babcock Manufacturing, which later became the Fire Extinguisher Manufacturing Company, and which was subsequently sold to American LaFrance in 1900, quickly became and was to remain the largest chemical engine manufacturer in the United States. By 1886, just fourteen years later, Babcock and its successor, Fire Extinguisher Manufacturing Co. had produced and placed more than two thousand chemical engines in active service across America.

But Col. Charles T. Holloway of Baltimore was not far behind. The first Chief Engineer of the Baltimore Fire Department, Holloway lost no time in forming a company to manufacture chemical engines and ladder trucks. He delivered the first Holloway chemical engine to Chicago in 1872, the same year Boston placed its Extinguisher Companies No. 2 and No. 3 in service. In 1873 Boston then purchased a chemical engine from each of the new manufacturers, one Babcock and one Holloway. Evidently Boston preferred the Babcock, as during the next seventeen years they purchased eleven more.

Becoming Babcock's close rival, Holloway produced the second largest number of chemical engines and also was bought out by American LaFrance in 1900. Essentially the Babcock and Holloway engines were similar, both makes being well-built to rigorous fire service standards, the only significant difference being the mechanism used to combine and mix the sulphuric acid and soda water.

There were three types of mechanisms: "The Champion," "The Champion-Babcock," and "The Holloway."

"The Champion"

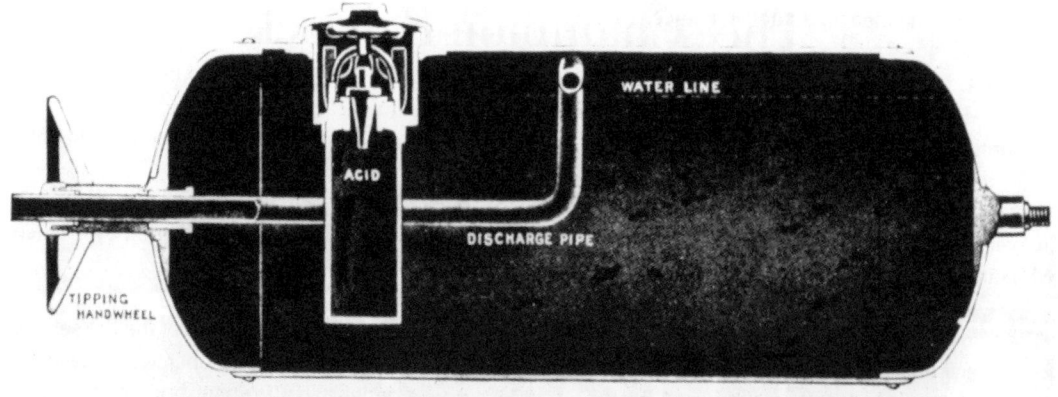

"THE CHAMPION" TANK
Sectional View

"Champion" Tank — Remove acid jar, unscrew stopper cage and remove. The function of the cage is to prevent the stopper from falling into the tank. Owing to great strength in this part being unnecessary and to space limitations, the cage is lightly constructed. Care must be exercised in handling, so that damage will not result.

Fill jar to bottom of glass screw in neck with 66-degree sulphuric acid. Mix up required amount of soda in bucket nearly filled with water. Stir contents until soda is dissolved and then pour into tank. Next add sufficient clean water to fill chemical tank to lower edge of acid jar supporting cage. By filling the tank after dissolved soda has been put in, an even better mixture is secured. Replace acid jar in tank. Screw on cap, being sure that gasket is in place and that cap is tight enough to prevent any leaks at this point.

To discharge tank, it is only necessary to remove the safety pin and turn tank over by means of the wheel provided; pressure will immediately generate. Before turning over, be sure that all valves are closed except that leading to hose.

By rocking the tank occasionally during discharge a better mixture and better working pressure is obtained.

When the tank has become discharged, bring it into the upright position. Before entirely removing the tank cap be sure that all pressure has been exhausted. Even a pressure seemingly insufficient to register on the gauge may blow the cap out of your hands were it not allowed to escape through the vent provided. Half way up the thread on the cap a small hole is drilled to the outside edge. When the cap is partially unscrewed this hole is uncovered and it allows any remaining gases to escape before the cap is removed. Therefore, partially unscrew the cap and allow pressure to exhaust itself before attempting the removal.

Be careful not to allow any of the raw acid or soda to reach the painted or other finished surfaces of the apparatus when filling the chemical tank. Either sulphuric acid or bicarbonate of soda in their raw state are harmful to finished surfaces if allowed to stay on them. When either of these ingredients is dropped on the paint, immediate spraying with clean cold water is essential. DO NOT RUB IT OFF.

If a sponge or chamois skin is used too quickly the paint will be marred. When the apparatus is sprayed with the mixed solution it should be removed as quickly as practical. If the chemical is properly mixed, any far-reaching harm will not result, other than would result from muddy water being allowed to dry on the paint. Both are to be avoided.

The Champion-Babcock

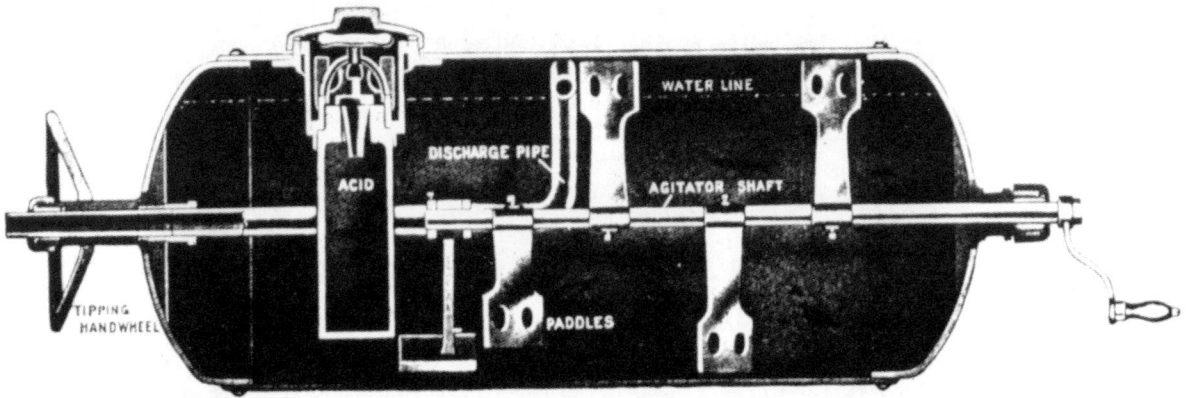

CHAMPION-BABCOCK TANK
Sectional View

"Champion-Babcock" Tank — This tank is operated practically the same as the "Champion." It, however, has the additional feature of an agitator. This agitator is a series of paddles attached to a central shaft which is rotated from the operating side of the tank by means of a lever. The rotation of the shaft causes the paddles to agitate the contents of the tank, which more readily unite. Higher and quicker attainment of pressure is thereby secured.

"The Holloway"

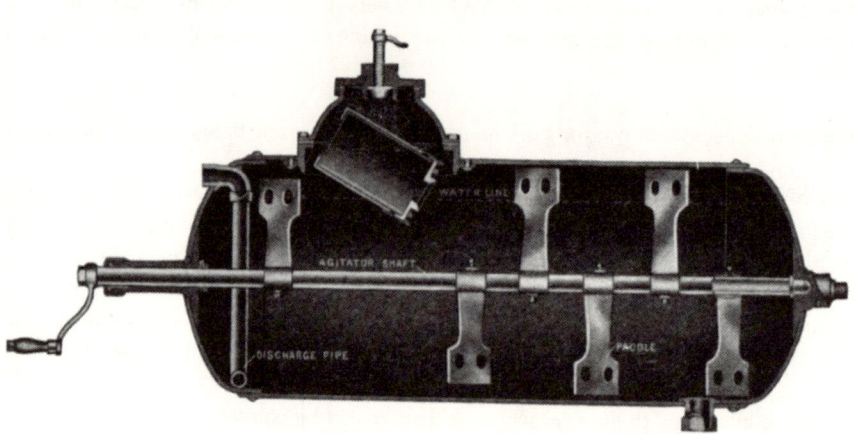

Sectional View. Holloway Tank.

"Holloway" Tank — This tank is readily identified by the hood which surmounts the tank at the filling port. The principle of operation in this type is somewhat different from those already described. In place of the tank revolving in order to dump the acid jar and begin generating as in the other types mentioned, the tank is of the stationary type. The acid jar is turned over, for discharging, by a small lever just outside the tank hood.

To charge the Holloway tank, follow the instructions as to putting in the mixed soda and then the water. Fill to the bottom of acid jar cage as in other types, then fill the acid jar and place in cage. The stopper for the jar is incorporated with the tank top cap. When acid jar has been placed in the tank be careful that it does not turn over. Back off stopper screw until this is up against bottom of cap.

Screw on cap and exercise usual precautions as to gasket and tightness of cap. Then carefully screw down stopper screw. This stopper is tapered to allow easy entry of the stopper into the jar. Also make a tight joint against acid leakage. When the stopper is just about to seat, move jar dumping lever gently backward and forward for half an inch or so to facilitate the stopper reaching a proper seat.

To discharge tank, simply raise stopper from bottle by means of screw and dump bottle with the lever provided.

All chemical tanks must be protected against freezing in cold weather.

When coiling hose into the chemical basket, it should be coiled in a clockwise direction looking down upon the top of the basket. This will have a tendency to tighten the chemical hose coupling when pulling it from the basket. If hose is coiled in the opposite direction it will likely result in uncoupling the hose and losing the contents of the chemical tank.

As time went on, dozens of new chemical engine manufacturers, both large and small, many regional, jumped on the band wagon. Three of the largest competitors to Babcock and Holloway were Seagrave, Peter Pirsch, and Obenchain-Boyer of Logansport, Indiana. In the Obenchain-Boyer chemical tank shown below, a plunger on the glass bottle was struck by a mallet inside the tank, literally breaking the glass. The pieces of glass were caught in the bottle cage. A hand crank then turned a spiral screw type agitator to more thoroughly mix the chemicals.

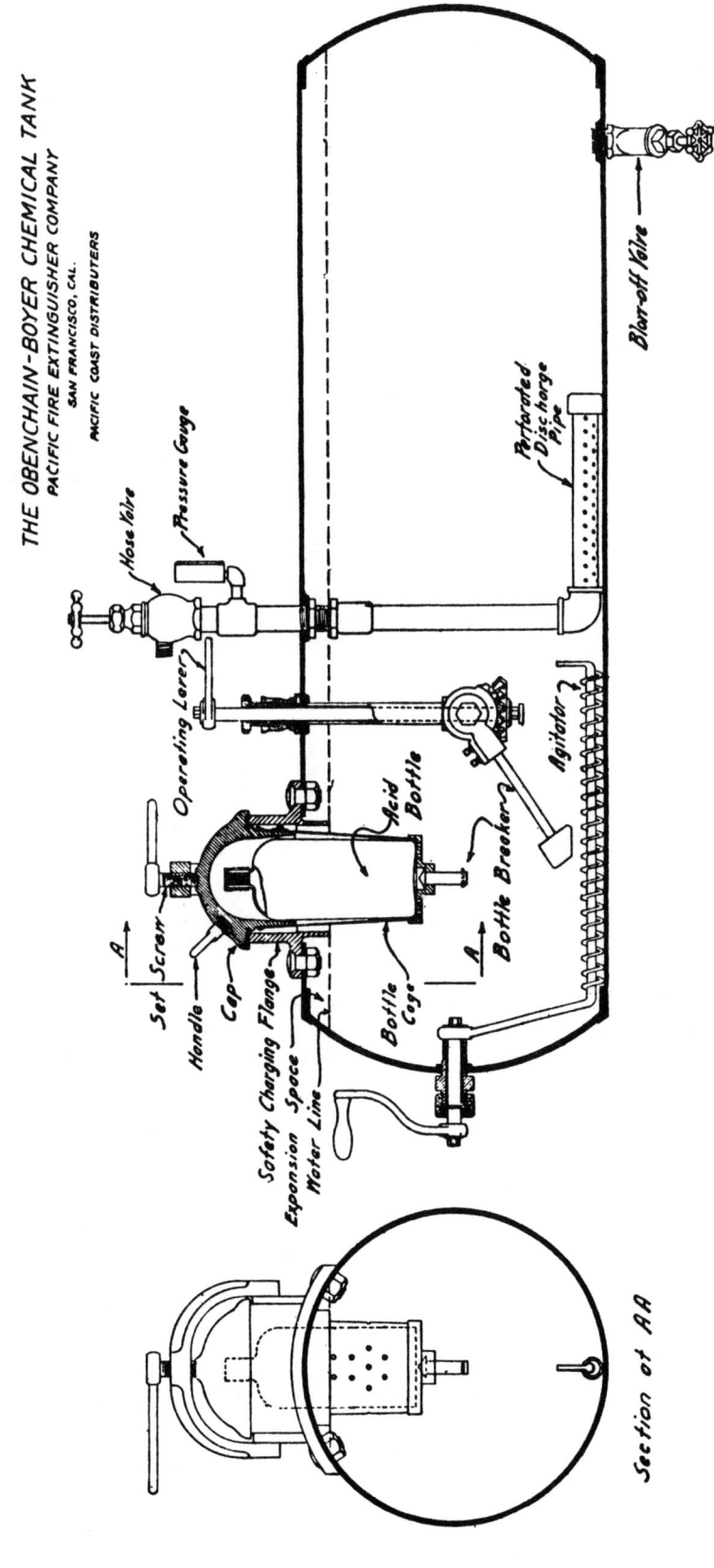

15

The OBENCHAIN-BOYER CHEMICAL ENGINE

IN designing and constructing a Chemical Fire Engine, the fact which must be kept in mind is that it is a machine which is to be used under the most trying circumstances, when the operators will be working under a strain of excitement, when any hitch or delay will prove fatal to the purpose which the engine is designed to serve, that whenever it is called upon, it must, without fail, instantly and vigorously respond, and that, from the nature of the service, it will be subject to the severest kind of usage.

The six points to be looked for in passing judgment on a Chemical Fire Engine are as follows:
Absolute sureness of action.
Speed and ease of operation.
Provision for dissolving the soda.
Provision for sealing the acid.
Security against overcharging.
Grade of material used, standard of workmanship, general appearance and finish.

With the above points in mind we invite you to the consideration of the Obenchain-Boyer Chemical Fire Engine. Reference to the accompanying drawing of the interior of the tank will render these points discussed clear.

OPERATION OF ENGINE

When this type of engine is drawn to the scene of a fire there is but one thing necessary to set it in operation, that is, to pull, through a quarter turn, the operating lever on top of the tank. This lever actuates the bottle-breaking mechanism and within ten seconds from the time it is operated the pressure gauge will register from 140 to 180 pounds pressure.

BOTTLE BREAKING DEVICE

The device for breaking the bottle is so simple that it will, in all cases, positively effect its purpose. The bottle holder is so constructed that it retains all the broken glass and not a particle of it can fall into the tank. (See drawing).

OPENING *and* CLOSING TANK

WHEN the engine has discharged itself, it is of the utmost importance that it be capable of being easily and quickly recharged. The cap, which gives access to the tank and holds the acid receptacle, is held in place by means of a hinged arch which permits of its being very much more quickly and easily removed than if secured by any other means, especially so if a threaded cap were used, as is the case with most chemical engines. Where a threaded cap is used, it requires the use of a long lever wrench and several turns of the cap before it can be removed. Also, in replacing this style of cap, difficulty is often experienced in starting the thread.

Threaded cap is not easy to start under the most favorable circumstances, and, if in the inevitable excitement of a fire, the thread should be started wrong and forced on, it would be ruined and possibly put the engine out of commission at once.

RECHARGING

WHEN the cap is removed on the Obenchain-Boyer engine, which is done by loosening the hand-led set-screw on top, throwing the arch over and lifting out the cap by means of the attached handle, the water and soda must be put into the tank, and it is necessary that this be accomplished quickly and at the same time, that the soda should be thoroughly dissolved in the water.

All types of this engine are provided with a flexible agitator for thoroughly dissolving the soda. This agitator consists of a spiral coil hung on a vibrating shaft, and when operated it comes in close contact with the shell of the tank, stirring each grain and particle of the soda up into the water and thoroughly mixing and dissolving it. No chemical engine will work at its highest efficiency unless this is accomplished, and in this type of engine is the only agitator to accomplish it without an expenditure of more time and labor than can reasonably be expected of any fire company in the hurry and excitement of recharging a tank at a fire.

With this agitator the engine can be recharged just as quickly as the water can be emptied into it. As soon as the soda and a few buckets of water have been emptied into the engine, one man should begin operating the agitator and continue it until the tank is filled. This will assure the complete mixing and dissolving of the soda in the water and it will be ready to receive and combine with the acid the instant it is released from the bottle.

The sealed acid bottle, in the meantime having been placed in the bottle cage—an operation which requires not more than five seconds—it is then necessary only to replace the cap, throw up the arch, and screw down the set screw, when the engine is again ready for service.

SAFETY CHARGING FLANGE

A VERY important advantage in this engine is the means of preventing the hasty or inexperienced operator from overcharging the tank, that is, charging it so full of water that a dangerously high pressure will be generated. In charging the tank, if it is filled entirely full of water, any ordinary soda and acid charge, when the water is at a temperature of about 70 degrees Fahrenheit, will create a pressure of over 1,000 pounds per square inch. The only means whereby this dangerously high pressure can be avoided is to leave an expansion space above the water. (See drawing).

It is to be noted in this connection that most manufacturers of chemical engines direct that their tanks be filled only to within three or four inches of the top; this to prevent over-charging. In the Obenchain-Boyer engines, however, no directions are needed to prevent the over-charging, the engine being so constructed that it is absolutely impossible to fill the air space at the top of the tank. This is done by means of a Safety Charging Flange which extends down several inches into the tank and provides air space so that even when a tank is filled to overflowing, this air space, extending the full length of the tank, is maintained. This method is far superior to the overflow valve in general use for this purpose, which, in the haste and excitement of a fire, is often forgotten and left closed.

SEALED ACID CONTAINER

THE acid is held in a glass bottle hermetically sealed. This method has three distinct advantages: It prevents the deterioration of either the acid or the soda; it prevents the escape of the acid fumes which would corrode the metal of the tank above the water line; it prevents the slopping over of the acid during the run to a fire. Frequently, in machines with loose stoppered acid receptacles, the acid will slop over during a run to such an extent that the engine will have to be recharged whether other conditions render this necessary or not.

To insure positive action in a chemical engine with loose stoppered acid receptacle it is necessary to recharge the tank at least once each year, but with the acid receptacle hermetically sealed, one charge will remain in perfect condition for a great many years. This fact alone insures perfect action of the engine even though it be neglected during a long period.

TANK CONSTRUCTION

THE tanks of the Obenchain-Boyer engine are made of the highest grade of open hearth flange steel of a tensile strength of 90,000 pounds per square inch. General opinion has, in the past, leaned towards copper tanks in preference to steel, but this is an erroneous belief, however, as substantiated by the following facts. The best hammered copper has a tensile strength of only 30,000 pounds per square inch, and, at the same time, weighs approximately 12½% more than steel for equal bulks. This makes a copper tank of equal strength with steel weigh about twenty-five per cent more than a steel tank. Further—for a long time it has erroneously been supposed that copper was less easily attacked than steel by the soda solution of a chemical engine, and this would be the case if the tank were filled with pure water. The solution, however, being rendered alkaline by the presence of the soda, becomes an absolute preventative of corrosion of iron or steel, while on the other hand, it has a pronounced effect on copper, oxyd carbonate of Brunswick green being formed, the copper wasting away in proportion. A steel tank, which stood charged with soda solution for thirty-two years, and which was discharged and recharged several times during that period, was cut into pieces and it was found that the metal had wasted less than 1-64th of an inch in this length of time, and this waste was almost entirely confined to that portion of the tank above the water line. That there was any wastage at all was caused by the fact that the acid receptacle used was one with a loose stopper and the straight deterioration shown was caused by the action of the acid fumes. These fumes always escape from the acid receptacle equipped with the stopper loose enough to open automatically when the receptacle is turned over, but this difficulty is entirely overcome by the use of the hermetically sealed bottle.

THE material and workmanship on this type of engine is of the very highest grade; the axles are made of the best axle steel with case-hardened spindles, and all the steel used in the tank construction is required to fulfill the condition of bending cold 180° double on itself without sign of crack or fracture on the outside of the bent portion.

The auxiliary equipment is of the very best grade obtainable and is such that it will fulfil all of the requirements of general fire duty in connection with hand-drawn apparatus. These engines are approved and labeled by the Underwirters' Laboratories.

CAP
Catalog No. A 1007

BOTTLE CAGE
Catalog No. A 1008

BOTTLE BREAKER

ACID BOTTLE
Catalog No. A 1006

The OBENCHAIN-BOYER CHAMPION TANKS

THESE tanks are made of the best grade cold rolled copper, being double riveted along the horizontal seam, with single riveted heads. Fittings are all made of best grade red brass and interior of tank and all fittings covered with thick coating of tin and lead to prevent corrosion. Tested to 350 lbs. These tanks are regularly furnished with plain heads, one acid bottle, but without piping, valves, or by-pass.

Catalog No. M1242

Interior Equipment of Tank

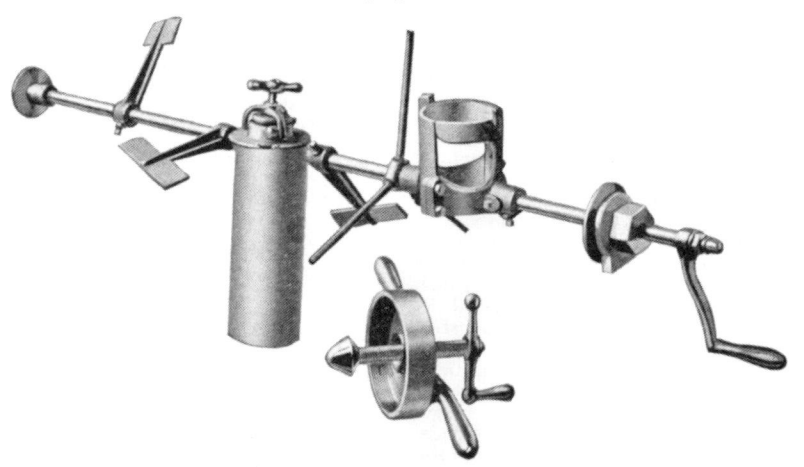

"PIRSCH" Chemical Tanks

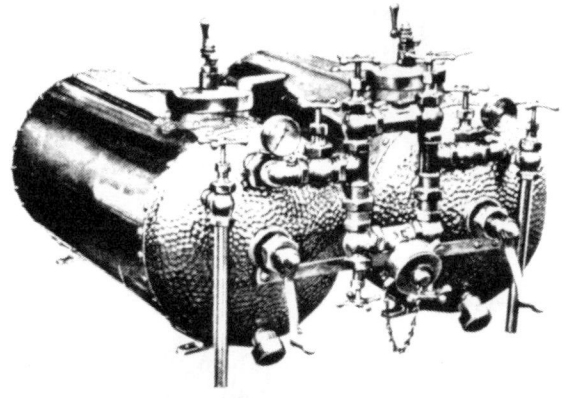

"Pirsch-Standard" Type Double Tank Arrangement
(Showing Double Piping System Used on all Styles)

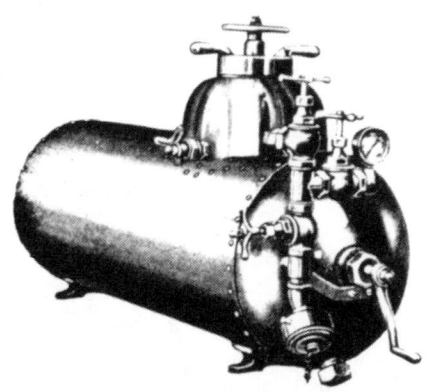

"Pirsch-Holloway" Type
Chemical Tank

"Pirsch" Chemical Tanks

"Pirsch" Champion Type Tank

paddles or agitators. A cork attached to screw stem running through filling cap with crank on top, holds the bottle and shaft from inverting and prevents any acid from spilling out of the bottle or any water from splashing into it. In operating, the handle on top of tank is screwed upward which releases the bottle, and the crank at end of tank is given one-half turn which tips the bottle, after a few seconds the crank is given several turns which mixes the contents of the tank. This tank is provided with a washout spud in the lower end of head for washing out any undissolved soda which remains in tank after using; it also has an overflow pipe with valve, which is a positive guard against overfilling the tank with water, either when refilling at a fire, or after a fire, in the engine house.

STEEL OR COPPER

We build tanks of either steel or copper whichever is preferred. The copper tanks, of course, cost more money than the steel. In our method of constructing the steel tanks and testing, the life is about as long as the copper. A copper tank looks nicer and has a greater intrinsic value, but the fire fighting efficiency of both is about the same. All castings and fittings inside as well as outside of tanks being of best grade red brass. The copper which is used in our copper tanks is heavier than is ordinarily used (35-gallon $\frac{1}{8}$-inch thick, 40-gallon $\frac{5}{32}$-inch thick). The steel shells for the steel tanks are seamless, drawn from steel rolled especially for this purpose and will stand a tensile test of 1100 pounds per square inch. Both copper and steel tanks are tested to 350 pounds hydrostatic pressure before leaving our factory, which is more than they receive in actual service. Our chemical tanks are in service in some of the largest fire departments in the world, such as New York, New Orleans, San Francisco, Akron, Duluth, Buffalo, etc., where they are receiving hard, persistent service.

COATING

Our chemical tanks, both steel and copper, are coated on the inside with a composition of tin and lead, which, from experience, has proven to be the most efficient in preventing corrosion of the metal itself. Examination of several tanks coated as above after several years of service has been made at our plant without any trace of corrosion being found.

BOTTLES

"Pirsch Standard" acid bottles are made with a lead tube lining, copper sheet jacket tinned with lead and tin on both sides, top of acid bottle fitted with brass castings. "Pirsch Champion" bottles have glass screw neck on which rests a loose lead stopple or cork. All bottles supplied with lifters, "Pirsch Champion" style being screwed to top casting to hold cork from falling out.

PIPING

All piping is made of best grade seamless brass tubing; all valves, elbows, tees and nipples of best grade red brass. Joints are threaded, sweated and soldered; the complete piping is then tested to 350 pounds hydrostatic pressure.

Single tank piping consists of three valves, one pressure gauge, one $2\frac{1}{2}$-inch filling connection with by-pass, cap and chain and one connection for attaching chemical hose.

Double tank piping consists of six valves, two pressure gauges, one $2\frac{1}{2}$-inch filling connection with by-pass, cap and chain and one connection for chemical hose. Piping arranged so that each tank operates independently of the other and the empty one may be recharged while other is discharging.

CAPACITIES

All types and styles built in 25, 30, 35, 40, 50 and 60 gallon sizes, in either single or double units.

FINISH

All copper tanks may be polished finish or nickel-plated. Steel tanks are painted any color desired, with piping, brass cap and all brass parts on exterior of tank either polished or nickel-plated. Steel tanks may be supplied with nickel plated ends at extra cost.

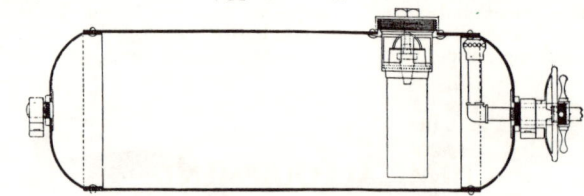

Interior View "Pirsch-Champion" Type Tank Showing Operating Mechanism

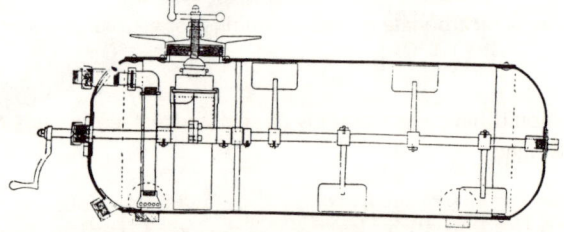

Interior View "Pirsch-Standard" Type Tank Showing Operating Mechanism

PETER PIRSCH & SONS CO.

"Pirsch" Chemical Tanks

"Pirsch" Standard Type Tank

Chemical tanks as manufactured by us represent the highest degree of development thus far attained in this line of product. Our experience in this line covering a period of more than twenty-five years has enabled us to go through the experimental stage years ago, and through keeping close watch on chemical tanks of our own as well as other makes in service, to overcome and eliminate any defects or faults as they presented themselves thus making our tanks today as nearly perfect as can be attained.

TYPES AND STYLES

We build chemical tanks of the "Pirsch-Champion," "Pirsch-Standard" and "Pirsch-Holloway" types, each tank efficient and capable of doing effective fire duty. The "Pirsch-Champion" style is of the loose stopple type, and while it is the most simple in operation and efficient for slow moving horse and hand-drawn vehicles, is not considered by us the most suitable for use on fast moving motor fire apparatus.

The "Pirsch-Holloway" tank is of the sealed bottle type and has agitator for mixing the contents. This type tank is used largely by some of our large fire departments. This tank requires more space for installation than the other types owing to the large dome on top.

The "Pirsch-Standard" type chemical tank we consider the logical tank for use on motor fire apparatus of all types, owing to its efficiency, compactness and simplicity. This "Pirsch-Standard" tank is sometimes known as the semi-Holloway type on account of the agitator and sealed bottle features. The acid bottle is carried in a cage on the same shaft which carries the

CHEMICAL EQUIPMENT

Proper Solutions — The generally recognized formula and one that secures the best results, is composed of 66-degree sulphuric acid, bicarbonate of soda, and water. These ingredients to be in the following ratio: One-fifth part acid and two-fifths parts bicarbonate of soda to number of gallons of water or capacity of tank.

For example, a 40-gallon tank should have eight pounds of acid and sixteen pounds of soda. Acid, although a liquid, is measured by weight.

Pressure and operating results are often jeopardized by the improper mixing of the soda and water. The best method, and one which should be followed, is to mix the soda by stirring it in a bucket of clean water before placing it in the tank. If this method is pursued, the soda will not "cake" in the bottom of the tank as it is thoroughly dissolved. This is a very essential operation for charging the tank at fire stations.

When it is necessary to charge the tank at a fire, this method must naturally be dispensed with and the soda dumped into the tank quickly. Even this will produce good working pressure as the charge is fresh and the soda has not had opportunity of forming a hard cake at the bottom of the tank. Better results may be obtained by rocking the tank slightly; this has a tendency to give a better mixture.

If called upon to extinguish a bonfire of some sort, be sure and play the stream at the base of the fire. The chemical fumes thus rise through the flames and smother it. There is a limit of effectiveness of a chemical stream in the open. If the fire is too large and composed of very inflammable material, some difficulty may doubtless be encountered; in a closed room however, the chemical stream is highly efficient.

A fire can often be extinguished, although it has assumed large proportions, with a minimum of water damage, as the stream used is very small in comparison with that of the fire pump.

The three types of tanks used on American-LaFrance apparatus are the "Champion," which is our standard, and the "Champion-Babcock" and "Holloway" which are only furnished where specified. Each of these three types have special features, and directions for filling and maintenance are given below.

Chemical Tank Parts

Holloway Bottle Each..........$21.00

Standard Bottle Each............$21.00

Pirsch - Champion Bottle. Each $20.00

Acid Receptacle Canister
Pol. fin.....$10.00
N. p. fin... 11.00

Champion Parts

Tipping lever........$3.00   Valve stem handle $ 1.50   Screw................. $3.00

Lifter..$2.00   Glass neck..$1.00   Lead stopper..$1.00   Lifter............$6.00

Soda Canister
Pol. fin................$11.00
N. p. fin.............. 12.00

Single Chemical Tank Piping

Single Piping, with Valves, 2½" Gauge,
2½" Hose Connection and By-Pass
Price, pol. fin....$65.00   Price, n. p. fin....$70.00

Double Chemical Tank Piping

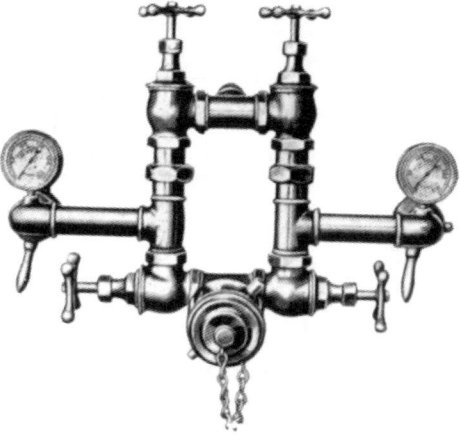

Double Piping with Valves, 2½" Pressure
Gauge, 2½" Hose Connection and By-Pass
Price, pol. finish..................................$110.00
Price, n. p. finish..................................$115.00

Plate 167

Plate 168

SODA

We use the very best quality of bicarbonate of soda obtainable. This is made to our order, and we use it in large quantities; consequently, always have a supply of fresh soda on hand. The good results obtainable from a chemical fire extinguisher or chemical fire engine depend largely upon the use of good chemicals.

Our soda is guaranteed to be the very best, and if bought in barrels containing
400 pounds, the price is..6c. per pound
In 112-pound kegs, price is..6½c. per pound

ACID

We use the very best grade of 66-degree Sulphuric Acid (Oil of Vitriol). It is guaranteed to be full strength, and it sells by the carboy, containing about 180 pounds, at 7c. per pound.

Principal Dimensions
Champion & Champion-Babcock
Acid Jars of Non-Corrosive Metal.

Cylinders	C	D
25 & 30 Gal.	$7\frac{3}{4}$	$3\frac{7}{8}$
35 & 40 Gal.	$10\frac{11}{16}$	$3\frac{7}{8}$
45 & 50 Gal.	$11\frac{1}{4}$	$4\frac{1}{4}$
55 & 60 Gal.	$13\frac{11}{16}$	$4\frac{1}{4}$
80 Gal.	$10\frac{1}{4}$	6
100 Gal.	12	6

Plate 2148b

Principal Dimensions
Holloway
Acid Jars of Non-Corrosive Metal.

Cylinders	B	D
25 & 30 Gal.	$5\frac{5}{16}$	$4\frac{5}{16}$
35 & 40 Gal.	$7\frac{9}{16}$	$4\frac{5}{16}$
50 Gal.	$8\frac{1}{8}$	5
60 Gal.	$8\frac{11}{16}$	$5\frac{7}{16}$

Plate 2148a

CHAMPION ACID JAR CAGES

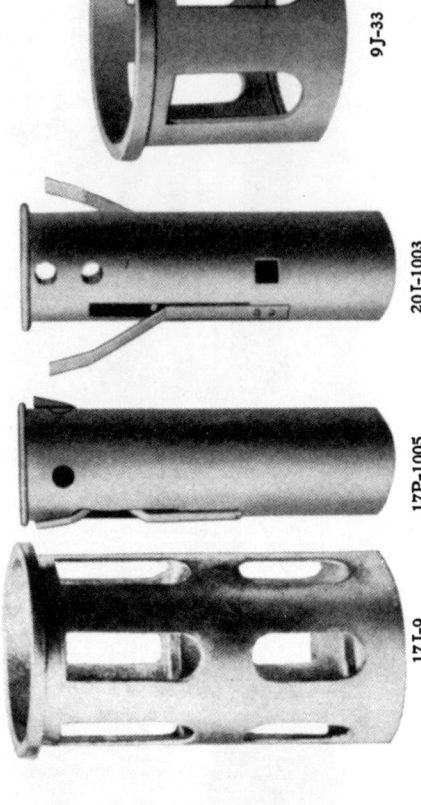

17J-9   17P-1005   20J-1003   9J-33

HOLLOWAY ACID JARS

Plate 137A
HOLLOWAY Jar Only

Plate 137
Holloway Jar with lifter

4N1030	25-30 gal. each		$22.00
4N1014	35-40 gal. each		24.00
4N1015	50 gal. each		26.00
4N1025	60 gal. each		37.00

Plate 137A
Holloway Jar Only

25-30 gal. each		$15.00
35-40 gal. each		17.00
50 gal. each		19.00
60 gal. each		27.00

NOTE Holloway type jars for tanks do not require lifters. Lifters are required for extra jars and are furnished with non-corrosive valves.

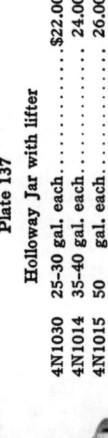
Plate 137
HOLLOWAY with Lifter

PNEUMATIC ACID SIPHON   CARBOY INCLINATOR

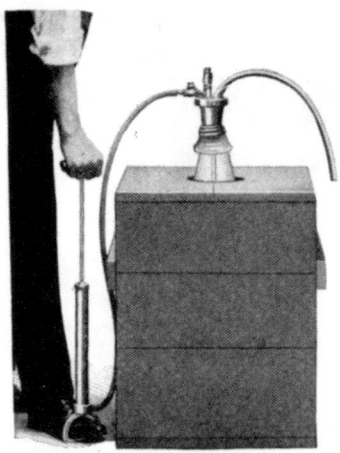

Plate 170

Plate 171

Acid always under control. No splashing. No danger.
Just a few strokes of the pump, and the acid starts to flow; an occasional stroke keeps it flowing smoothly without squirt or splash.
Price, nickel finish, complete with pump..$10.00

Is easily clamped to the carboy and insures the quick handling of its contents with safety.
Price.................................$15.00

ACID JARS AND BOTTLES

Champion Jar with lifter and stopper

2N1011	25 gal. each	$15.00
2N1012	35-40 gal. each	18.00
2N1013	45-50 gal. each	18.00
2N1014	55-60 gal. each	20.00
2N1053	80 gal. each	25.00
2N1031	100 gal. each	30.00

Champion Jar Only

25 gal. each	$11.00
35-40 gal. each	14.00
45-50 gal. each	14.00
55-60 gal. each	16.00
80 gal. each	21.00
100 gal. each	26.00

17J1003

Plate 136
CHAMPION
With Lifter
and Stopper

Plate 136A
CHAMPION
Jar Only

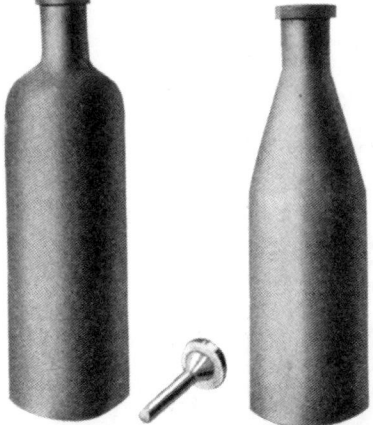

17P23   20J1002

Acid Jar
For No. 7 Labelled Engine
17J1003 Each.....................$20.00

Lead Acid Bottle
For No. 20 Labelled Engine
20J1002 Each..........$8.00

Lead Acid Bottle
For No. 20 Unlabelled Engine
17P23 Each..........$8.00

CHEMICAL HOSE BASKET

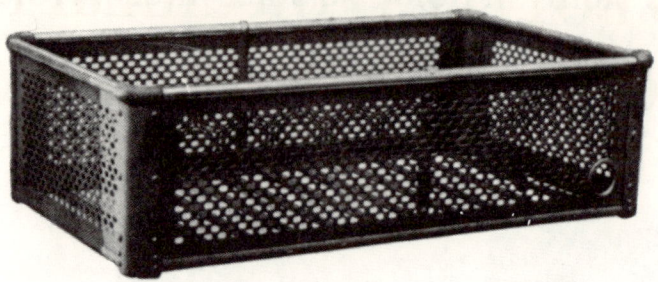

Plate 134

Capacity, 250 feet of ¾ inch or 200 feet of 1-inch chemical hose. Perforated steel panels, heavy frame and corners, brass rail at top, brass rings at port for hose connections. Finished in either polished brass or nickel plate. Specify which.

4 M1007 Hose basket, painted...$55.00
4 M1007 Hose basket, primed... 45.00

Dimensions: 23 inches wide, 42 inches long, 11½ inches deep.

Brackets for making basket swinging type. Price on application.

CHEMICAL HOSE REEL

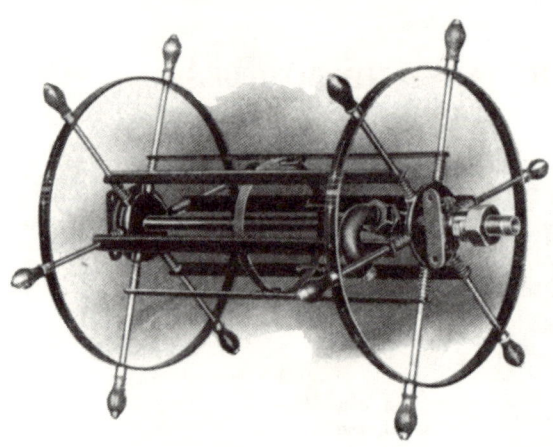

Plate 135

For chemical fire apparatus. Highest grade design and workmanship. Hollow axle, with elbow following natural curve of the hose.

Fitted with take-up springs under cross-slats. Brass piping and trimmings. Ready to attach to apparatus. Capacity, 250 feet ¾-inch chemical hose or 200 feet 1-inch chemical hose. Finished in either polished brass or nickel-plate. Specify which.

2 G1000 Hose reel, painted..$80.00
2 G1000 Hose reel, primed.. 70.00

ECCENTRIC CHEMICAL ENGINE NOZZLES

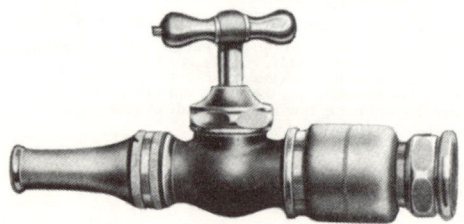

Plate 621

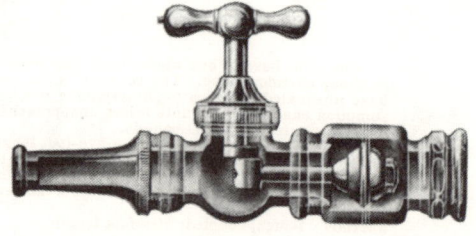

Plate 621

ACID BOTTLES FOR CHAMPION, BABCOCK AND HOLLOWAY ENGINES

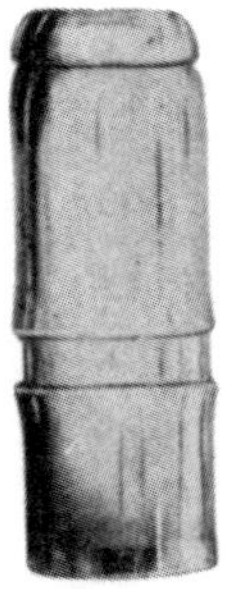

14P23—O-S-No. 1
Babcock
Size, 6¼ x 2⅛
Each.....$ .75
Per doz... 8.00

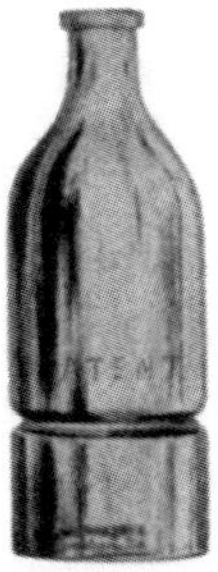

2P21—No. 1
Babcock
Size, 6 x 2¼
Per doz...$2.50
Per gross 18.50

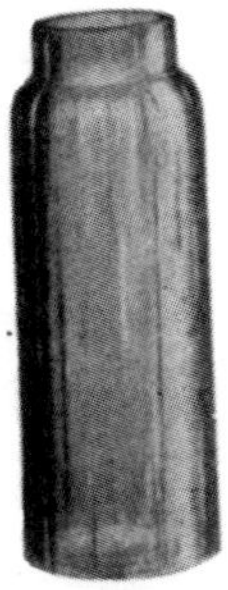

28P6
No. 1
Champion
Size, 6⅜ x 2⅜
Per doz...$6.00

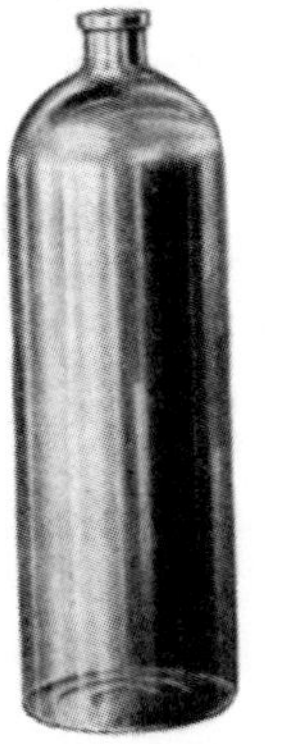

50 and 60 gal.
Holloway Eng.
Size, 12¼ x 4
Per doz..$9.00

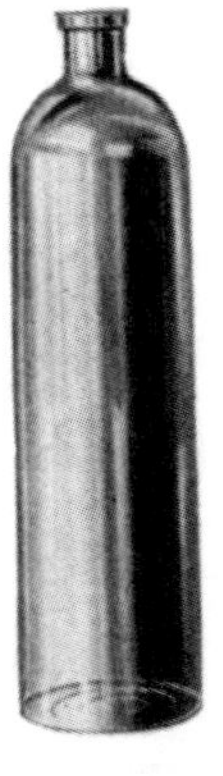

35 and 40 gal.
Holl'y Eng.
Size
12⅛ x $3\frac{7}{16}$
Per doz..$9.00

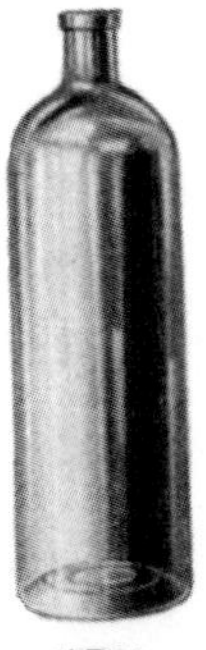

17P13
25 and 30 gal.
Holl'y Eng.
Size
$10\frac{3}{16}$ x $3\frac{3}{16}$
Per doz..$9.00

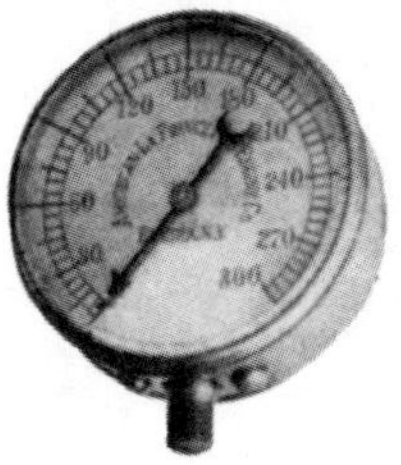

Plate 560

CHEMICAL ENGINE GAUGE

2V7. Price....$7.50

CHEMICAL HOSE

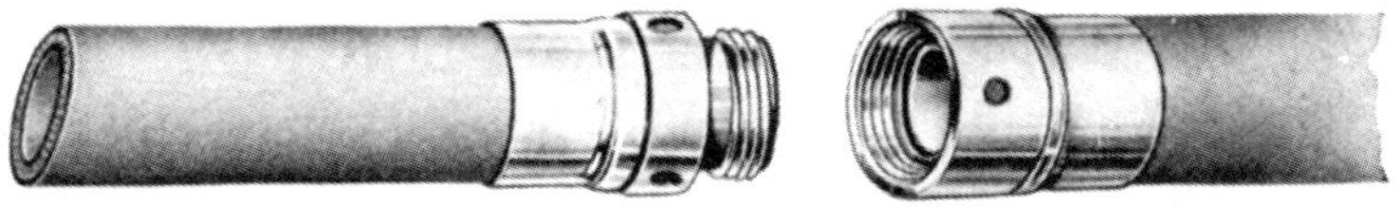

Plate 506

Chemical hose is one of the most expensive types of hose, if properly made. There is no economy in cheap chemical hose. The contrary is the case. We use over one hundred thousand feet a year and have unequaled facilities for determining the best article for the purpose. Furthermore, our great consumption enables us to quote prices unapproached by others, quality being considered. Made of four-ply heavy duck, tested to 500 pounds.

	½-inch, coupled, twenty-five-foot length	$17.00
3M268	¾-inch, heavy coupled, fifty-foot length	30.00
3M269	¾-inch, light, coupled, fifty-foot length	28.00
3M1131	¾-inch, coupled, Underwriter's labeled hose for two wheel chemical engines, fifty-foot length	30.00
3M1009	1-inch, coupled, fifty-foot length	45.00

Be sure to send records of threads so that correct threads can be cut on new couplings

II
THE HYPE THAT SOLD CHEMICAL ENGINES

AMERICAN-LaFRANCE
CHEMICAL *On Standard Ford Chassis*
Instant Response

THE safety of your own dear ones, your responsibility as a good citizen, demand that you get complete information about this wonderful motor fire-fighter at once. Be up-to-date and SAFE. "Chemical" has 40 times the fire-fighting efficiency of water. The outfit herewith is the product of the world's greatest manufacturer of fire apparatus.

THE SATURDAY EVENING POST
February 17, 1917

(This advertisement is continued on the next page)

(The top portion of this advertisement is shown on the preceeding page)

These American towns have already bought American-LaFrance Chemicals on Ford Chassis, and the list is incomplete:

Anthony, R. I.
Antioch, Cal.
Arcade, N. Y.
Arlington, Texas
Averyville, Ill.
Battle Mountain, Nev.
Berwick, Me.
Boothbay Harbor, Me.
Bowie, Texas
Brookville, Pa.
Buck Hill Falls, Pa.
Buxton, Me.
Caldwell, Texas
Campbell Hall, N. Y.
Carey, Ohio
Caribou, Me.
Carroll, Iowa
Chariton, Iowa
Cleveland, Okla.
Clinton, Wis.
Corcoran, Cal.
Cotuit, Mass.
Dalton, Mass.
Dewey, Okla.
Dunkirk, Ind.
Dwight, Ill.
Edinburg, Ind.
Effingham, Ill.
Galeton, Pa.
Girardville, Pa.
Gladbrook, Iowa
Gladstone, N. J.
Granger, Texas
Grayling, Mich.
Greenlake, Wis.
Greenville, Texas
Grosse Pointe Shores, Mich.
Hamburg, N. Y.
Hamilton, Mass.
Haskell, Texas
Hopewell, Va.
Hugo, Okla.
Interlaken, N. Y.
Kewaskum, Wis.
Kelley Island, Ohio
LaPorte, Ind.
Las Cruces, N. M.
Laurel Springs, N. J.
Littleton, Mass.
Lynnfield, Mass.
Mattoon, Ill.
Merchantville, N. J.
Mineola, Texas
Mt. Rainier, Md.
Munday, Texas
New Boston, Ohio
Palo Alto, Cal.
Pittsburg, Cal.
Pocatello, Idaho
Portsmouth, N. H.
Princeton, W. Va.
Reading, Pa.
Richfield Springs, N. Y.
Royse, Texas
Sandusky, Ohio
Shiner, Texas
Silvis, Ill.
Southampton, L. I.
Tekamah, Neb.
Vicksburg, Mich.
Warsaw, Ill.
Westfield, Pa.
West Long Branch, N. J.
Willow Grove, Pa.

—that feeling of SECURITY in an emergency

A lace curtain. A spark. A flame. A telephone. A woman, tremulous—yet trusting. The clanging bell in the town fire-house. The American-La France Chemical tearing along the road like a hound unleashed, snorting in encouraging response to the woman's S. O. S. The timely arrival, discrediting the fastest horse-drawn apparatus. The fire put out before it got under way. The woman's gratitude. The return to the fire-house.

Your turn may be next. Don't wait for the fire alarm, with consequences no man can foretell. Your town now, this very minute, needs the protection of this trusty, speedy, economical life-saving apparatus. Write at once for further information. Your turn may be next.

AMERICAN-LAFRANCE FIRE ENGINE COMPANY, INC.
THE LARGEST BUILDERS OF FIRE DEPARTMENT APPARATUS IN THE WORLD
Elmira, New York

The myth of the "chemicals" (nothing more than soda water propelled by carbon dioxide gas, a tiny fraction of which may have been suspended in the solution) persisted for as long as chemical engines were manufactured. Note that the following claims were made 43 years apart:

Babcock, 1874	"Carbonic acid gas is both the working and extinguishing agent, which, bulk for bulk, is 30 times as effective as water, the 200 gallons of the first class engine being equal to 6,000 gallons of water."
J.E. McWilliam & Son, Chemical Engine Manufacturer in Hubbardstown, Mass. 1904	"The Board of Underwriters claims that a solution of bi-carbonate of soda is 30 times more efficient than water in extinguishing fire."
American LaFrance 1917	"Chemical has 40 times the fire-fighting efficiency of water." (See ad on page 28)

Reprint from a 1918 Pacific Fire Extinguisher Co. Catalog
(West coast distributor for Obenchain-Boyer Chemical Engines)

Chemical Engines

CHEMICAL engines not only constitute a positive and comparatively inexpensive form of protection for small towns, isolated plants and private estates, but also comprise an important adjunct to effective fire protection in municipal fire departments and in industrial and commercial properties under city protection.

THE distinctive feature of this class of fire appliance lies particularly in the fact that they are light, take up little space, and are easily moved. In addition, they are always ready for instant service, can be readily handled by one or two persons, and are extremely simple in operation.

THE principle upon which chemical engines operate is extremely simple, and identical to that of the portable extinguisher of the soda-and-acid type: Sulphuric acid emptied into a solution of bi-carbonate of soda and water generates a large volume of carbonic acid gas. The pressure developed by this gas within the cylinder serves to expel the liquid contents of the tank under a heavy pressure—the natural extinguishing properties of the liquid being materially enhanced by the presence of the gas carried along with the stream as this gas, mixed with the air in a small proportion, tends toward preventing combustion. This gas is carried by drafts into places which are inaccessible to the stream itself, and adds to the extinguishing properties of the machine—the extinguishing properties of one gallon of chemical being rated as equal to 40 gallons of water.

THIS general type of equipment fills a distinctive and important field in fire protection. It would be rather absurd to advance the claim that the chemical engine meets the same general requirements demanded of automatic sprinkler installations, or the modern type of municipal fire apparatus—its particular advantage lies in its ability to promptly extinguish any kind of a fire, either within a building or without, before the fire becomes intrenched. Its use obviates water damage and the solution is effective on materials and substances not extinguishable by the use of water.

MAINTENANCE costs are practically negligible as the engine is mounted on wheels and only a comparatively small amount of hose is required, which is specially prepared and will last for years with reasonable care. The chemicals with which the engines are charged are both commercial commodities which may be readily obtained in any drug store and it is unnecessary to recharge a machine more than once a year, except if it is emptied for the purpose of extinguishing a fire.

IN presenting this line of equipment the Pacific Fire Extinguisher Company has been guided by one objective only—the merchandising of chemical engines which are quick in action, efficient, strong, durable and absolutely safe. The material and labor entering into their construction are of the best, the finishing is distinguished by the same grade of excellence, and the various models listed have been carefully chosen for the respective purposes for which they are best suited.

THEIR comparative low cost, their great effectiveness, and the fact that a large percentage of fires are put out exclusively by their use, either in the hands of municipal departments, or volunteer organizations, or by private brigades or individuals, commends them for extensive use in villages, industrial p l a n t s, commercial houses, warehouses, railroad properties and private estates.

Note the last sentence of the third paragraph, "…. the extingushing properties of one gallon of chemical being rated as equal to 40 gallons of water."

The manufacturers cannot be blamed for overexaggeration; fire chiefs themselves, from coast to coast, sang the praises of the fast-acting engines that put out 80% of their fires. The following chapter recounts the fire service enthusiasm for chemical engines.

III

THE AMAZING ACCEPTANCE OF CHEMICAL ENGINES

In 1895 the Fire Extinguisher Manufacturing Co. of Chicago went "fishing" for testimonial letters from Fire Chiefs who had Champion or Babcock Chemical Engines in their departments. Hundreds of replies flooded in singing the praises of Chemical Engines. Rather than reproducing some of the letters in their entirety, the author has exerpted some of the more interesting comments, which are listed in alphabetical order by cities:

ALEXANDRIA, MINN. For six years two No. 5 Champion single tank Chemical Engines were the only apparatus we had. They have always done good work. No department is complete without one.

ANGOLA, IND. At five recent fires the Champion Chemical Engine was of incalcuable value.

BALD MOUNTAIN, COLO. We have been using Champion Chemical Engines for five years and have put out about 90% of all fires, in fact every one that we have got to in any time at all.

BRAIDWOOD, ILL. We have two of your No. 5 Champion Chemical Engines and have used them for about fifteen years. They are out of sight.

BROCKTON, MASS. We now have three double 60 gallon engines, each drawn by two horses. Nearly 75% of our fires are extinguished by their use alone.

CANTON, OHIO A perfect piece of mechanism, neat and symmetrical in appearance and wonderful in power of execution.

CHATTANOOGA, TENN. I consider the Chemical the best piece of apparatus in the department.

CRESTON, IOWA Seven-eighths of our buildings are wood, yet the second building at the same time has got to burn yet, all on account of the Babcock.

DALLAS, TEX. The Champion Chemical Engine is so satisfactory that the city purchased the second one Monday last.

DETROIT, MICH. We have five of the Champion Chemical Engines. They put out 70% of our fires.

DEXTER, IOWA The cylinder turns upside down to mix the chemicals. It has done some noble work for us.

ELGIN, ILL. I would not think of running a fire department without one.

EVANSVILLE, IND. We have two Champion Chemical Engines which put out 75% of our fires. No fire department is complete without Chemical Engines.

EMMETSBURG, IOWA We have used a two wheel Chemical Engine for about fifteen years. Fully 75% of our fires have been extinguished with it.

FARIBAULT, MINN We have a No. 4 Champion double cylinder fifty gallon Chemical Engine. It is one of the greatest fire fighting apparatus ever built. I am ready at any time to defend its merits.

FINDLAY, OHIO Have used No. 3 Champion Chemical Engine for nine years. It is the thing to fight fires. Fire Departments should not be without one.

GALLATIN, TENN. We have two Champion Chemical Engines which have saved our city from destruction several times. They are worth their weight in gold.

GALVA, ILL. Although we have all modern apparatus, we could not be induced to give up our dear old Champion Chemical Engine.

HAVELOCK, NEB. Thus far have extinguished 100% of all fires with the Champion. We have no other fire protection and could not do without it.

HAVERHILL, MASS. Having run on a Chemical nine years as a Captain, I think they are the best apparatus for extinguishing fires.

JACKSON, MICH. Our department has not lost a dwelling house or store building since we obtained our first Chemical Engine four and one-half years ago.

INDIANAPOLIS, IND. Have used Champion Chemical Fire Engines twelve years. Fully 60% of our fires are extinguished by them.

KEOKUK, IOWA We are well pleased with the Champion double fifty, and would not part with it if we could not get another.

LAWRENCE, MASS. We have been using two of your Chemical Engines, and they have never gone back on us.

LEROY, N.Y. All departments who wish to be successful cannot afford to be without a Chemical. Get them to a fire early, and your success is assured.

MADISON, S.D. We have two No. 5 two wheel 100 gallon Champion Engines. I do not consider a fire department complete without a Chemical Engine.

MINNEAPOLIS, MINN. We have used Champion Chemical Engines for the past twelve years and have at the present time five. No department is properly equipped without having a number of Chemical Engines.

NORTHVILLE, MICH. Your two wheeled Champion Engine is just the thing.

OSWEGO, N.Y. We have been using an eighty gallon Chemical Engine for eight years and our Board has just purchased another one.

PARIS, TEXAS Would not do without it for twice what it cost us.

PRINCETON, ILL. We have had two Babcock Engines for twenty-five years, and are in good condition yet.

PROVIDENCE, R.I. We have two Babcocks and one Champion. I cannot understand how any city can afford to be without them.

QUANAH, TEXAS The No. 5 four wheeled Champion Chemical has proven to be one of the greatest fire "fighters" that it has ever been my pleasure to use. It has checked over 80% of all fires.

SABETHA, KANS. Our engine is a double fifty gallon Babcock. She's about 15 or 16 years old, but she's all right. It knocks fire just like breaking sticks.

SEATTLE, WASH. We use three Chemical Engines in this city, two Holloways and one Champion. The two Holloways put out about 10% of the fires and the Champion about 60%. I wish we had all Champion Chemicals in this city. The one we have here is worth its weight in gold.

VERMONTVILLE, MICH. Your Champion two wheeled Chemical Engine is a dandy. It can't be beat.

WEST NEWTON, MASS. Our No. 4 Champion Chemical Engine has proven itself to be the most valuable piece of apparatus in the department.

WINNIPEG, MAN., CANADA The three Champion Chemical Engines have ever been a reliable safeguard over the lives and property of our citizens.

WETMORE, KANSAS Our Champion Chemical Engine has saved our town two or three times.

WICHITA FALLS, TEXAS We use the Champion Chemical and feel we could not well be without it. It has extinguished about 85% of the fires.

WILSON, KANS. Since our Champion Chemical Engine has been in our use we have had several fires. In all we kept the fire confined to the building where it started except once in a store where sixteen gallons of gasoline exploded.

(Well, you can't win 'em all.) This just a sampling of the hundreds of testimonial letters that were received. Fire Chiefs nationwide obviously thought the Chemical Engines were the greatest thing since sliced bread. Even if "Chemicals" weren't 30 to 40 times as effective as water, this misunderstanding can easily be excused since the engines got to the fires and got them out while they were still small. Whoever invented today's fast attack mini pumpers reinvented the wheel . . . they've been around for more than a century!

During September 1886 Captain E.F. Martin of the Boston Fire Department addressed the Massachusetts State Firemen's Convention on the subject, "CHEMICAL ENGINES: THEIR EFFICIENCY AND BENEFITS". Here are some extracts from his speech:

It was Shakespeare who said that
"A little fire is quickly trodden out,
Which being suffered. rivers cannot quench."

I have yet to learn that Shakespeare contemplated the introduction of chemical engines at the time he wrote those lines.

My experience with chemicals was of several years' duration, and during the interval I attended many fires, over sixty per cent. of which were extinguished by chemical engines without assistance from the other portion of the department called.

* * * *

In extinguishing fires with a chemical engine the principal advantage lies in being able to do so with little water damage, and, as they are lighter, they can be taken to the place of action much quicker than our heavy steamers. Let it be understood that I do not recommend a chemical engine to take the place of a whole fire department, *but I do claim that they will accomplish more in the line of fire extinguishment than they have ever received credit for.*

* * * *

No good general or other military officer would bring a battery to bear on a picket line of the enemy while he had sharpshooters and infantry at his command. Just so in the extinguishment of fires; no good and experienced chief engineer, his assistant or company officer, should place a water tower, steam fire engine or hydrant stream in service when he has a chemical engine that will do the work required. The chemical engine will accomplish a great deal, if only given a chance. I remember where they paid for themselves at one fire. I might give you dates and losses, but for these you probably do not care. As I understand it, what this convention wants to know are facts, and these I will attempt to give to the best of my judgment.

* * * *

They have demonstrated their usefulness in no uncertain terms in combating with oil fires. I believe, from personal observation, that a stream from one of these machines, through a quarter inch nozzle, will do more execution than one from an inch nozzle of a steam fire engine, in this special line, where floors are covered and saturated with oil and other inflammable liquids. *They are the best of guardians in tenement house districts ; in fact, I might say they are the poor man's friend.*

I have seen these machines prove their efficiency when brought into action on the roofs of buildings which were in danger from a fire in progress near them. As spark chasers and annihilators they have no equal in the line of fire apparatus. Again, they can be used to a great advantage in buildings adjoining the one on fire, especially where there is machinery with open belt-holes. They also receive credit from every quarter for their usefulness in the extinguishment of brush and grass fires.

* * * *

It seems to me that a question of vital importance to your organization is the consideration of fire department equipments, *with a special eye cast on the introduction of chemical engines in every city and town that is willing to provide themselves in the line of fire protection.*

Less than a year later, in August 1887, Col. Charles Pickett of the Baltimore County Maryland Fire Department delivered a paper entitled, "THE VALUE OF CHEMICAL FIRE ENGINES" to the Virginia Firemen's Association, as follows:

"Among the more recent and effective instruments which have come into use as 'fire apparatus' is the 'chemical engine.' It is a well known fact that any atmosphere containing over 5 per cent of carbonic acid cannot sustain combustion. The chemical engine is the practical application of this principle. The apparatus consists of strong metal vessels, containing alkaline solutions, into which, by a simple mechanical contrivance, sulphuric acid is precipitated. The reaction is instantaneous. Carbonic acid gas is freed, and the pressure thus exerted makes the water absorb many times its volume of gas, and on the opening of a cock or valve this same pressure ejects a stream on the fire, carrying with it in suspension this gas. On reaching the fire, it at once becomes liberated, and being heavier than air, so to speak, overflies the fire and thus stops combustion.

"Owing to the facts stated above, you can readily see with what rapidity the machine can be brought into service. No time is required to lay a long line of hose or to make water connections. Being charged, the machine is always ready and one, say of a capacity of 100 gallons, equals in extinguishing power 4000 gallons of water, and it can be recharged in a very brief space of time, as extra charges are always carried with it. Thus, where the water supply is limited, its efficiency is beyond conception. The best steamer made is of no avail at a fire if there be no ample supply of water.

"Enough to carry a chemical engine a whole day would be consumed by a steamer in less than an hour's work. When fires occur in stock of goods, or where damage by water would be great, one of its chief advantages is demonstrated, since only one-fortieth the amount of water is used. Reports show that in this very salvage one has been known to pay for itself in a single fire. The hose used is so small and light that one man can easily carry a hundred feet of it, either into a building or up a ladder, whereas it takes several to handle our regular hose. In oil fires, it is the only thing that will bring them under control, as water seems but to add fuel to the flames. I could quote you here the success in this line at several oil refineries near Baltimore and at other points, when the ordinary apparatus completely failed and the chemical engines subdued them but do not desire to consume so much of your time."

If Col. Pickett had been on the scene in Wilson, Kansas, perhaps he could have even extinguished the fire fed by sixteen barrels of gasoline. Chemical Engines were the darlings of fire departments large and small, north, south, east and west.

But insurance interests in Louisville, Kentucky evidently needed still further evidence of the merit of chemical engines since they sent letters to fire chiefs throughout the country asking (1) how long have you been using Chemical Fire Engines? (2) What percent of your fires do you extinguish with these engines? (3) can you recommend Chemical Fire Engines? Once again the testimonial letters flooded in. Here are the tabulations:

City & State	How many years in use?	% of Fires Extinguished?	Can You Recommend Chemical Engines?
Addyston, Ohio	1 year	90%	Have paid for themselves many times over.
Alexandria, Minn.	12 years	80%	No department is complete without one.
Arcade, N.Y.	20 years	90%	Yes.
Arlington, Mass.	6 years	70%	Yes.
Atlanta, Ga.	8 years	33%	Yes.
Bald Mountain, Colo.	5 years	90%	Could not do without it.
Bath, Me.	10 years	50%	Yes.

City & State	How many years in use?	% of Fires Extinguished?	Can You Recommend Chemical Engines?
Battle Creek, Mich.	5 years	90%	No department complete without them.
Binghamton, N.Y.	6 years	50%	Yes, sir.
Birmingham, Ala.	7 years	75%	They are invaluable.
Bloomington, Ill.	9 years	50%	No department equipped without them.
Blythebourne, N.Y.	5 years	50%	Yes.
Boise City, Idaho	6 years	75%	I do, most emphatically.
Boston, Mass.	21 years	10%	I consider them a great auxiliary to any department.
Braceville, Ill.	12 years	70%	Would not dispense with the Chemical Engine.
Braidwood, Ill.	15 years	50%	For quick and effective work, they can't be beat.
Brandon, Man.	3 years	50%	Yes.
Brockton, Mass.	8 years	75%	I certainly can.
Buffalo, N.Y.	19 years	45%	I can.
Burlington, Vt.	7 years	75%	I can, and think they are just the thing.
Cambridge, Mass.	15 years	Large	No Department fully equipped without one.
Canton, Ill.	2 years	33%	Would not be without one.
Canton, Ohio	8 years	40%	No Department equipped without them.
Carthage, Ill.	10 years	50%	Filled all requirements. No repairs. Always ready for instant service.
Cassleton, N.D.	12 years	65%	Most decidedly, yes.
Cedar Bluffs, Neb.	3 years	100%	Have given good satisfaction.
Chattanooga, Tenn.	4 years	70%	No Department complete without Chemical.
Chehalis, Wash.	4 years	40%	I can.
Chicago, Ill.	25 years		Yes. Worked at 1,610 fires in 1894.
Chicago Heights, Ill.	3 years	90%	Would not part with the Chemical Engines at any price.
Chicopee, Mass.	15 years	90%	Yes, I am a firm believer in them.
Clarinda, Ia.	22 years	80%	Yes.

City & State	How many years in use?	% of Fires Extinguished?	Can You Recommend Chemical Engines?
Clark, S.D.	8 years	90%	They are world-beaters.
Cleves, Ohio	5 years	60%	No Department should be without them.
Colorado Springs, Colo.	12 years	80%	I can.
Columbus, Ga.	22 years		Yes, we do good work with it.
Columbus, Ohio	16 years	50%	Yes, after 16 years' experience.
Council Bluffs, Iowa	4 years	75%	I certainly can.
Covington, Ky.	14 years	80%	Yes.
Cuthbert, Ga.	18 years	60%	Yes.
Dallas, Texas	7 years	50%	Very necessary auxiliaries to a Department.
Dayton, Ohio	15 years	50%	Yes.
Detroit, Minn.		67%	Yes.
Detroit, Mich.	19 years	50%	No Department should be without them.
Denver, Colo.	8 years	60%	Yes, they are one of our most valuable arms of the service.
Des Moines, Ia.	12 years	65%	Yes, would not do without them.
Dexter, Ia.	15 years	85%	Yes.
Donaldsonville, La.	19 years	40%	I can.
Dubuque, Ia.	5 years	70%	I do, most heartily.
Duluth, Minn.	18 years	60%	Yes.
Elgin, Ill.	12 years	50%	I can.
Elmira, N.Y.	10 years	60%	Yes, I can and do.
El Paso, Ill.	17 years	50%	I can.
Emmettsburg, Ia.	15 years	75%	Yes.
Evansville, Ind.	14 years	65%	No Department complete without them.
Evergreen Park, Ill.	2 years	100%	Does splendid service.
Findlay, Ohio	9 years	65%	Every time.
Fort Worth, Texas	3 years	40%	Could not do without them.

City & State	How many years in use?	% of Fires Extinguished?	Can You Recommend Chemical Engines?
Galva, Ill.	16 years	85%	Most assuredly.
Grand Island, Neb.	15 years	50%	Cannot speak too highly of them.
Halifax, N.S.	3 years	70%	No up-to-date Department should be without them.
Hamilton, Ill.	1 year	75%	No Fire Department complete without them.
Havelock, Neb.	2 years	100%	Could not do without it.
Haverhill, Mass.	18 years	75%	I think every place should have them.
Healdsburg, Cal.	12 years	50%	I most heartily recommend Chemical Engines.
Henrietta, Texas	4 years	50%	Consider nothing better.
Honolulu, Hawaii	8 years	40%	I unqualifiedly recommend the Chemical Engine.
Howell, Mich.	20 years	90%	Yes.
Ilwaco, Wash.	3 years	90%	Is worth its money. Has done good service.
Indianapolis, Ind.	12 years	60%	No Department complete without them.
Jackson, Mich.	11 years	70%	No Fire Department complete without them.
Jersey City, N.J.	23 years	80%	Yes, every time.
Joliet, Ill.	15 years	50%	Yes.
Kalamazoo, Mich.	10 years	50%	I can, very strongly.
Keokuk, Iowa	2 years	75%	I can, most heartily.
Kewanee, Ill.	20 years	50%	Yes.
Lafayette, Ind.	12 years	80%	Yes, would not do without them.
Lawrence, Mass.	15 years	50%	Yes, no Department equipped without them.
Le Roy, N.Y.	12 years	33%	Most certainly.
Lima, N.Y.	19 years	80%	Yes.
Lincoln, Neb.	5 years	55%	Could not dispense with them.
Long Branch, N.J.	21 years	Don't know	Yes.
Los Angeles, Cal.	3 years	60%	Yes.

City & State	How many years in use?	% of Fires Extinguished?	Can You Recommend Chemical Engines?
Lowell, Mass.	10 years	67%	Yes.
Lynn, Mass.	15 years	75%	Yes.
Madison, S.D.	12 years	80%	Yes, sir.
Manchester, N.H.	10 years	40%	Yes, 365 days in the year.
Marlin, Texas	2 years		I consider no Department complete without one.
Memphis, Tenn.	13 years	75%	Would not do without them.
Merrill, Wis.	10 years		Could not get along without them.
Milwaukee, Wis.	18 years	40%	No city is well provided without them.
Minneapolis, Minn.	22 years	60%	No Department properly equipped without them.
Monson, Mass.	8 years	70%	Yes.
Mt. Jackson, Ind.	3 years	60%	No village can afford to be without them.
Muncie, Ind.	18 years	65%	Yes.
Nashua, N.H.	8 years	70%	Yes, could not get along without them.
Newark, N.Y.	8 years	100%	Yes.
New Orleans, La.	20 years	Large	I consider no Department complete without them.
Newton, Mass.	10 years	50%	Yes.
Ogden, Utah	6 years	40%	Yes, by all means.
Omaha, Neb.	10 years	50%	Would not do without them.
Oswego, N.Y.	8 years	40%	I can, highly.
Owensboro, Ky.	5 years	70%	Yes, could not do without ours.
Paterson, N.J.	5 years	70%	Yes.
Pawtucket, R.I.	8 years	75%	Every time.
Peoria, Ill.	20 years	40%	Yes, can't see how any City can get along without them.

City & State	How many years in use:	% of Fires Extinguished?	Can You Recommend Chemical Engines?
Piqua, Ohio	3 years	60%	Yes.
Pleasant Hill, Ohio	8 years	75%	We can recommend them as very efficient.
Portage La Prairie, Man.	2 years	50%	Yes.
Portland, Ore.	3 years	90%	I can.
Portland, Ind.	7 years	75%	Yes.
Providence, Mass.	6 years	25%	Most assuredly yes.
Pueblo, Colo.	3 years	60%	Would not do without them.
Quanah, Texas	4 years	80%	I can, most heartily.
Quincy, Ill.	15 years	75%	Yes, and no Fire Department is complete without them.
Rochester, N.Y.	22 years	50%	Yes.
Rock Valley, Ia.	3 years	75%	Yes.
Rutherford, N.J.	5 years	40%	Yes.
San Antonio, Texas	3 years	50%	The greatest machine in the world for a Fire Department.
San Francisco, Cal.	5 years	50%	I would certainly say that Chemical Engines are a valuable auxiliary.
Schuyler, Neb.	5 years	80%	Yes; they are always ready.
Seattle, Wash.	4 years	60%	My experience that no City is well provided without them.
Sewickley, Pa.	3 years	50%	Yes, sir.
Sigourney, Ia.	1 year	80%	I can, beyond a doubt.
Sioux City, Ia.	5 years	55%	Yes, sir.
Somerville, Mass.	12 years	50%	Yes, sir.
South Bend, Ind.	9 years	10%	I can.
Spokane, Wash.	6 years	50%	No Fire Department is well equipped without them.
Springfield, Ill.	8 years	75%	Yes, in highest terms.
Springfield, Ohio	21 years	50%	Yes.
St. Joseph, Mo.	5 years	25%	Yes.

City & State	How many years in use?	% of Fires Extinguished?	Can You Recommend Chemical Engines?
St. Louis, Mo.	20 years	Large	Are first-class.
Stockton, Cal.	21 years	72%	No Department properly equipped without them.
Streator, Ill.	16 years	80%	Yes.
Syracuse, N.Y.	15 years	30%	No Department complete without them.
Terre Haute, Ind.	9 years	70%	By all means, yes.
Topeka, Kan.	22 years	65%	Would not be without them.
Tuskegee, Ala.	15 years		Yes.
Vincennes, Ind.	6 years	50%	Yes; every Department should have one or more.
Waltham, Mass.	7 years	65%	Yes.
Warsaw, Ill.	2 years	75%	I think they are the best protection we have.
Washington C.H., Ohio	12 years	50%	Yes, I can.
Wetmore, Kan.	6 years		Would not be without our Chemical Engine.
White Sulphur Springs, Mont.	7 years	95%	Yes.
Wichita, Kan.	5 years	50%	I can.
Wilson, Kan.	10 years	95%	Yes.
Winnipeg, Man.	14 years	70%	Decidedly yes.
Youngstown, Ohio	5 years	67½%	Yes sir.

Extract from Report of New York's Fire Commissioner:

"The value of these chemicals in fire extinguishing, according to the testimony of the Fire Department authorities of Boston, Baltimore, Philadelphia, Chicago, St. Louis, Buffalo, and twenty other leading cities of whom inquiries have been made, is to the effect that they are the most valuable of recent aids and expedients in arresting fire, their operations covering an absolute control of from 25 per cent. to 80 per cent. of all the fires in those particular cities."

IV

THE ALL-CHEMICAL ENGINE FIRE DEPARTMENT

Most fire departments in larger cities and communities share a similar history. Personnel-wise they began as volunteers, adding a paid driver or chief, then evolving into part-paid and part-volunteer, and finally all-paid. Equipment-wise the usual progression was from bucket brigade to hand pumpers to steamers to motorized apparatus.

Can you imagine a brand new fire department starting from scratch in 1881 with an all paid force of seven men operating seven new engines, all of which were *chemical engines?* From 1881 until the first steamers were purchased thirteen years later, the Baltimore County (Maryland) Fire Department had purchased a total of twenty-three chemical engines! They had no hand engines and no steamers. Chemical engines and ladder wagons comprised the entire apparatus roster. Even after the first steamers were purchased in 1894, six more chemical engines were added.

Could this unusual all-chemical engine fire department have been influenced in any way by its first fire chief, whose name was Charles T. Holloway, who happened to be one of the two leading manufacturers of chemical fire engines in the United States?

Charles T. Holloway, who was destined to become Baltimore City's first paid fire chief, was President of the Hope Junior Fire Company at age fifteen. While carrying on his father's watch and clock-making business, which he inherited, he orgainzed the first hook and ladder company in Baltimore, the Pioneer. He served as Pioneer's President from 1850 to 1859, at which time he was appointed Fire Chief of the newly organized paid Baltimore Fire Department.

CHAS. T. HOLLOWAY, First Chief Engineer, B. C

In 1864 he resigned as chief. But in 1868 he accepted the appointment to the position of Fire Inspector. Here he became responsible for new laws providing for the inspection of illuminating oils, and "prevention of the sale of such as are dangerous." He also organized the Salvage Corps, which remained in existence until 1958.

In 1870 he assisted in the formation of the paid Fire Department in Pittsburgh, Pa., and in that same year he formed the Holloway Manufacturing Co., which was to become one of the two largest manufacturers of chemical fire engines in the United States.

An early account of Col. Holloway's career states, "He is largely engaged in the manufacture of chemical fire extinguishers of his own invention, which he has brought very close to perfection." The tanks were made of copper, with the upper part polished. The engines had platform springs, oil tempered, and were constructed of Swedish steel. The apparatus had Archibald wheels varnished in the natural color. They had iron frames arched for the front wheels to turn under the frame or gally for carrying hose pipes.

The pump (for refilling the chemical tanks while at a fire, from any convenient water source) was located on the rear of the frame, and the apparatus also carried a suction sleeve (for drafting to refill the tank), one signal lamp, two hand lanterns, one hook, one pick, two buckets, a drivers seat, and a tongue for two horses.

On June 8, 1881, the Baltimore County Commissioners placed an order with the Holloway Manufacturing Co. for six double tank chemical engines costing a total of $12,800.00. But there was a catch. In order to obtain this order Holloway was to accept, and he was to remain the unpaid chief of the new paid department for two years.

Seven chemical engines were actually ordered, one for each of seven new fire stations. Each engine was drawn by two horses. "In case of a general alarm, all the engines will be required to come to the fire."

One of the first alarms was to William Kemper's beer brewery. The new firemen "moved bravely on, never halting to think of their wives and dear little ones whom they had left at home, and who might, per chance in a few hours be bereft of their kind and loving fathers." This heroism confined damage to the brewery to a mere $50.00.

During the thirteen year period from 1881 until 1894, the year the first steamers were purchased, the Baltimore County Fire Department bought an incredible 23 Holloway chemical engines. And after the purchase of the first two steamers, six more Holloway chemicals were added for a total of twenty-nine!

No other fire department in the nation operated that many chemical engines, much less to the exclusion of hand engines and steamers. In Baltimore County, chemical engines were not just a first attack force - they were the *entire* attack force for thirteen years!

CHARLES T. HOLLOWAY,

MANUFACTURER OF

CHEMICAL FIRE APPARATUS

COMPRISING

4 Wheel Double Tank Chemical Engines, 50 to 200 gal. capacity,
2 and 3 Wheel Single Tank Chemical Engines,
25 to 100 gallon capacity.

V
HAND DRAWN CHEMICAL FIRE ENGINES

Two-wheel hand drawn chemical engines were manufactured for more than half a century and were to be found in towns and villages all across America.

These engines had a tongue with handles, making their overall length from eight to eleven feet. Two men could get the engine to the fire if the distance was short and the way level. But for long runs and hills, the drag rope (50 feet of manila rope on a reel) was pulled out, and many willing hands pulled the engine.

One hundred feet of chemical hose, usually ¾" in diameter, was coiled in a hose basket, but some engines had a hose reel. One of the hand drawn chemical engines in the author's collection has a solid brass reel. Most of these engines carried lanterns, axes, pry bar, and a tool box with assorted spanners and wrenches, which included a wheel wrench. Many were elaborately striped and decorated. The chemical tanks were often copper with brass fittings, and they gleamed with frequent polishing. Some models had a gong which was struc k automatically by each revolution of the wheel. One or two brass containers for extra acid were usually mounted, which added to the dazzling appearance of these copper and brass beauties.

The following pages show drawings and photographs of hand drawn chemical engines made by some of the leading manufacturers.

HOLLOWAY
TWO-WHEEL
Single Tank Chemical Engine,

For Towns, Villages, Depots, Factories, &c., &c.

No. 1, 100 Gallons Capacity.   No. 2, 80 Gallons Capacity.   No. 3, 50 Gallons Capacity.

EQUIPMENT.—One tank, 150 feet 4-ply rubber hose, 1 hose pipe, 2 soda holders, 2 acid holders, drag rope, 1 axe, 1 pick, 2 lanterns, 1 spanner, wrenches, &c.

HOLLOWAY CHEMICAL ENGINE,

FOR VILLAGES, DEPOTS, MILLS, ETC.

No. 6, Capacity, - - - - - 50 Gallons.

DESCRIPTION OF ENGINE.

To be drawn by hand. Built and equipped as follows: Two Copper Tanks, each 25 gallons capacity. Tanks constructed as described in No. 2 Upper part of Tank polished. All Brass Work polished. Brass Moulding around the centre of Tanks at the platform. Iron Frame, arched for front wheels to turn under. Back Springs Elliptic. Front arranged as shown in cut. Patent Wheels, front wheels 3 feet high, back wheels 3 feet 6 inches high. Walnut Box on front for extra charges. Hose Gallery, 200 feet ½ inch 4 ply Rubber Hose. Hand Tongue, Hose Pipes, Wrenches, etc. Built of best material and handsomely finished.

No. 3 Chemical Engine, for Cities and Towns.

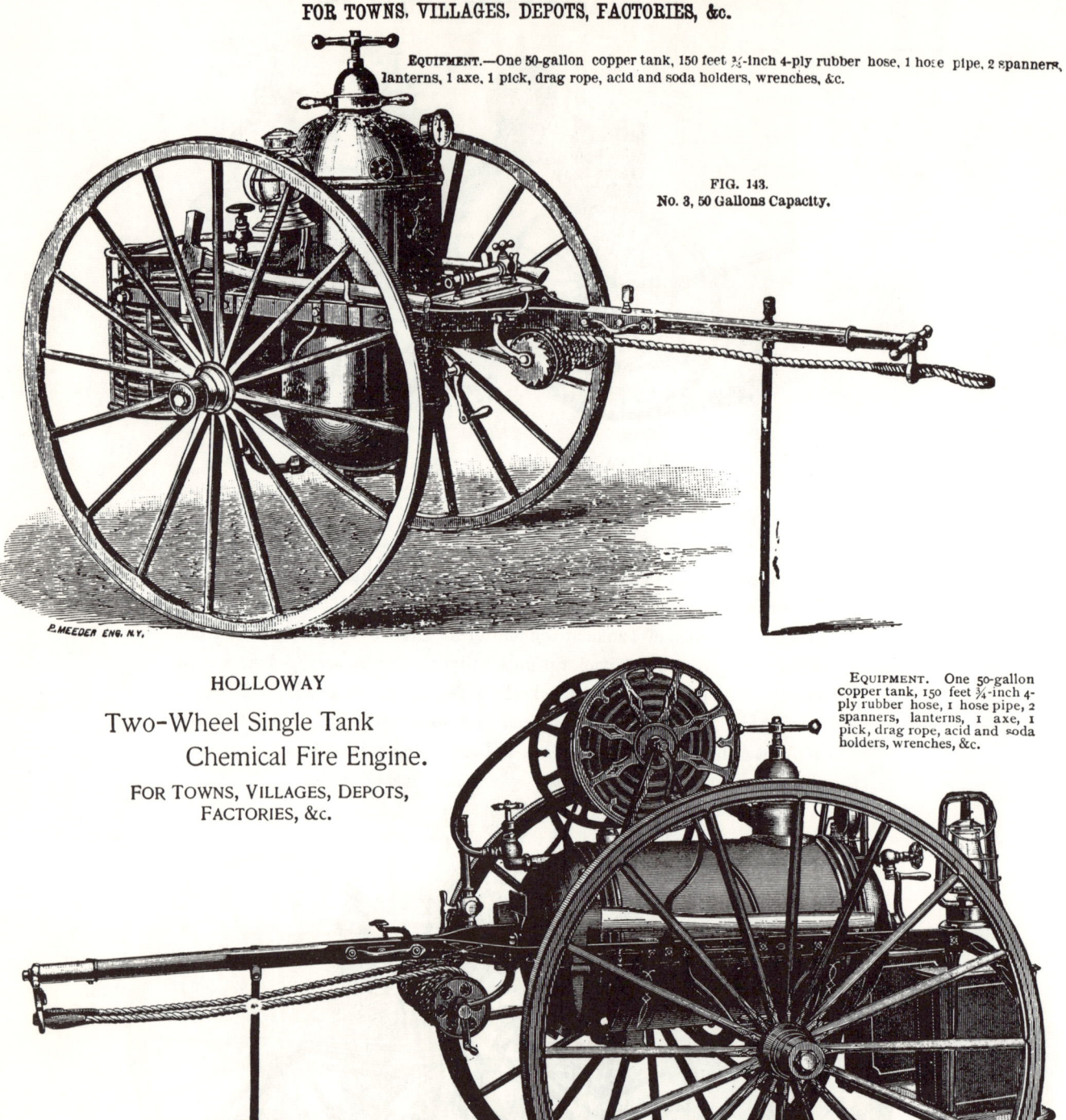

HOLLOWAY
TWO-WHEEL SINGLE TANK CHEMICAL FIRE ENGINE.
FOR TOWNS, VILLAGES, DEPOTS, FACTORIES, &c.

EQUIPMENT.—One 50-gallon copper tank, 150 feet ¾-inch 4-ply rubber hose, 1 hose pipe, 2 spanners, lanterns, 1 axe, 1 pick, drag rope, acid and soda holders, wrenches, &c.

FIG. 143.
No. 3, 50 Gallons Capacity.

HOLLOWAY
Two-Wheel Single Tank Chemical Fire Engine.
FOR TOWNS, VILLAGES, DEPOTS, FACTORIES, &c.

EQUIPMENT. One 50-gallon copper tank, 150 feet ¾-inch 4-ply rubber hose, 1 hose pipe, 2 spanners, lanterns, 1 axe, 1 pick, drag rope, acid and soda holders, wrenches, &c.

Single Tank, Three Wheel Chemical Engine.

51

DOUBLE 35-GALLON CYLINDER FOUR-WHEEL "CHAMPION" CHEMICAL ENGINE.
(To be Drawn by Hand.)

No. 5—FOUR-WHEEL "CHAMPION" CHEMICAL ENGINE (to pull by hand).

No. 6—TWO-WHEEL "CHAMPION" CHEMICAL ENGINE (to pull by hand).

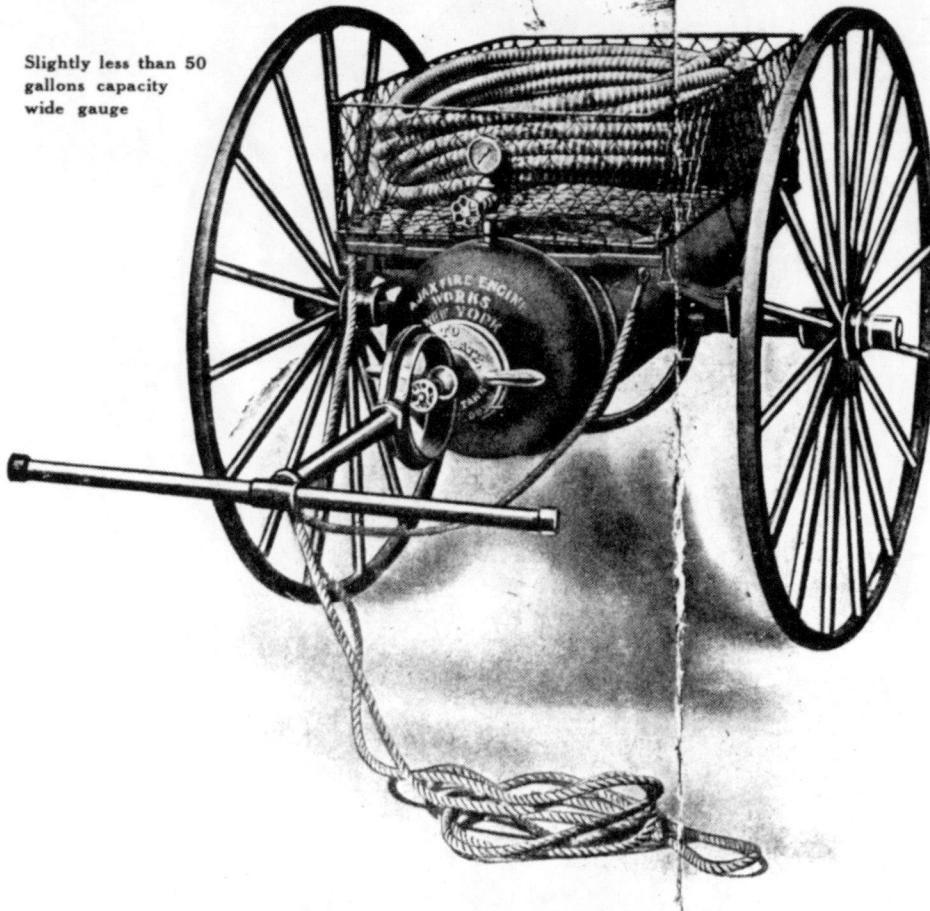

Construction and Equipment, Type 3 AJAX Chemical Fire Engine

Slightly less than 50 gallons capacity wide gauge

Construction and Equipment, Type 4 AJAX Chemical Fire Engine

TWO TANKS with 35 gallons capacity each

TYPE THREE.

The type 3 AJAX has an extreme width from hub to hub of 54 inches, and a tread of 44 inches, and is equipped with axles and springs.

These features prevent the possibility of the machine overturning when going around sharp corners, at a rapid rate of speed, and make it possible to attach the hand-pole to the back of a wagon or automobile, if desired, thus making it a "trailer" if occasion demands it. The latter is not practical with a narrow tread engine equipped with ordinary axles.

The type 3 AJAX is equipped with 100 feet of 7-ply chemical fire hose, and shut-off nozzle, the hose being wire armored which prevents "kinking." The hose is contained in an exceptionally large basket which allows it to be in large coils, thus insuring long life for the hose. There is also room in the basket for other fire equipment if desired.

The machine is equipped with an iron hand-pole and rope reel, so that several men can rapidly haul the machine. There is also a shut-off valve on the tank, and a pressure gauge, and we ship three charges of chemicals with each machine. (Additional charges can be secured from us or any drug store at slight cost.)

Tank:—Each cylinder is drawn cold, and pressed into shape out of a plate of special selected open hearth flange steel, without weld, rivet or seam. As this process affords the only reliable means of detecting hidden laminations or seams, the pressed steel tank is the only absolutely safe and secure reservoir for high pressure requirements. Each tank is tested at the factory to 400 pounds hydrostatic pressure, and it could actually stand a much higher pressure if it were necessary.

Capacity:— The capacity of the tank is nearly 50 gallons, which is equivalent to thousands of gallons of ordinary water in fire fighting effectiveness.

Acid Receptacle and Method of Operation:— The acid receptacle is constructed of desilverized lead, so as to be acid-proof and the same is protected from external injury by a covering of brass. A glass stopper of round marble shape, is in the neck of the acid receptacle, and is held in place by means of a valve to which it is attached and which clamps it down tight. This prevents the acid from accidentally spilling while the machine is being hauled, as the receptacle is sealed, which is desirable in machines for Fire Department use, or for any Customer who would probably haul the machine a long distance before the fire is reached. In order to operate the machine, this valve (which comes through the head of the tank) is first given a turn, which draws up the stopper out of the receptacle, and then the hand-pole of the tank is rested on the ground or floor. When the hand-pole is in this position, and the valve is open, the acid then flows out into the soda water solution and generates the pressure, which instantly produces a powerful stream which is thrown 75 to 80 feet.

TYPE FOUR

This engine is built especially for Fire Department service, although sometimes purchased by large industrial corporations or railroads. It has two 35 gallon tanks, seamless drawn steel, and is mounted on two wheels, regular artillery type, with iron hubs, brass hub caps, round edge steel tires, etc.

The frame is of forged steel, the axle is cranked. Best oil tempered half elliptic springs are used.

The hose is carried in a large hose basket, and consists of 100 feet of ¾ inch special chemical fire hose, with special shut-off nozzle.

We also give with this machine four acid receptacles, two brass acid receptacle holders, in rear of engine, two Fire Department lanterns and holders, one Fire Department axe with holder, one Fire Department crow bar with holder, one pressure gauge, one special forged hand-pole, fitted with pole grab, and hand holes, and in fact the machine is complete in all respects. Every thing is included with this machine that is contained on the Automobile Chemical Fire Engines, used by large City Fire Departments. Three charges of chemicals for each tank goes with the machine.

To operate the type 4 AJAX, it is only necessary to turn one of the operating wheels at the rear. This will revolve the tank on its hollow trunnions. A loose lead stopper drops from the acid jar, and permits the flow of acid into the alkaline fluid. The tanks are so arranged that one can be charged while the other is being discharged, and in this way you can have enough capacity to last as long as desired. It takes only a minute or two to recharge a tank. Nothing could be simpler, or more complete in all respects, for Town Fire Department use.

Single Tank, Three Wheel Chemical Engine.

J. E. McWILLIAM & SON

One 15 gallon and two 5 gallon tanks for a total of 25 gallons of "chemicals" are carried on this two wheel chemical "truck," which was "handsomely painted red, yellow, and dark green."

Fig. 13.—Two Wheel Truck in drawing position.

Notice Tanks hanging vertical, regardless of the position of the carriage.

American-LaFrance Fire Fighting Equipment

Plate 834
No. 9 Chemical Fire Engine

SPECIFICATIONS

No. 9. Forty-five-Gallon Champion Chemical Fire Engine
Operated by Revolving Cylinder on Trunnions

Cylinder—One (1), horizontal, 45 gallons capacity, Champion, seamless drawn steel, mounted on two wheels.
Frame—Forged steel.
Wheels—Best grade, with hub caps.
Tires—Round-edge steel.
Tongue—With drop prop.
Handle—Wheel type, in rear for inverting cylinder.
Hose—50 feet of ¾-inch, special four-ply rubber chemical hose, with brass couplings attached.
Nozzle—One (1) brass shut-off, eccentric, with rubber valve, chemical engine standard, two tips.
Basket—Wire, for carrying hose, of ample capacity.
Valve—One (1) (angle) of heavy brass.
Acid Receptacle—One (1), of noncorrosive metal.
Soda Bag—One (1).
Tool Box—in front.
Hose Spanner—One (1).
Wrench—One (1) axle nut. One (1) cylinder cap.
Painting—English vermilion, striped.
Capacity—45 gallons.
Height—Over all, 54¾ inches.
Width—Over all, 46 inches.
Length—8 feet.
Gauge or Track—38¾ inches.

Price, $500.00

American-LaFrance Fire Fighting Equipment

Plate 835
No. 10 Chemical Fire Engine

The author owns this engine, which has been restored to museum condition.

SPECIFICATIONS

No. 10. Forty-five-Gallon Champion Chemical Fire Engine
Operated by Revolving Cylinder on Trunnions

Cylinder—One (1), horizontal, Champion, seamless drawn steel, mounted on two wheels, capacity 45 gallons.
Frame—Forged steel.
Wheels—Best grade, with hub caps.
Tires—Round-edge steel.
Pole—With drop prop.
Handle—Wheel type, in rear for inverting cylinder.
Rope Reel and Drag Rope—Reel and 50 feet of drag rope, manila.
Hose—100 feet of ¾-inch, special four-ply rubber chemical hose, with brass couplings attached.
Nozzle—One (1) brass shut-off, eccentric, with rubber valve, chemical engine standard.
Basket—Wire, for carrying hose, of ample capacity.
Valve—One (1), angle, heavy brass.
Receptacles—Three (3), noncorrosive metal.
Receptacle Holders—Two (2), in rear of engine.
Soda Bags—Two (2).
Tool Box—In front.
Hose Spanners—Two (2).
Wrenches—All necessary wrenches.
Lanterns—Two (2), Fire Department style, with holders.
Ax and Holder—One (1), Fire Department style, with holders.
Crowbar and Holder—One (1), Fire Department style, with holders.
Gong—One (1) 10" automatic wheel striking, attached.
Painting—English vermilion, striped.
Capacity—45 gallons.
Height—Over all, 54¾ inches.
Width—Over all, 46 inches.
Length—8 feet.
Gauge or Track—38¾ inches.

Price, $650.00

American-LaFrance Fire Fighting Equipment

Plate 837
No. 11 Chemical Fire Engine—Rear View

SPECIFICATIONS
No. 11. Champion Chemical Fire Engine
Two cylinders. Capacity, 35 Gallons Each
Operated by Revolving Cylinders on Trunnions

Cylinders—Two (2), 35 gallons capacity each, Champion, seamless drawn steel, mounted on two wheels.
Frame—Forged steel.
Wheels—Best grade.
Hub Caps—Polished brass, closed ends.
Tires—Round-edge steel.
Axle—Cranked.
Springs—Best oil-tempered springs, half elliptic.
Hand Pole—Forged, fitted with pole crab and hand holds.
Handle—Wheel type, in rear for inverting cylinders.
Drag Rope—50 feet manila.
Hose—100 feet of ¾-inch, special four-ply rubber chemical hose, with heavy brass couplings attached.
Nozzle—One (1) brass shut-off nozzle, eccentric, with rubber valve.
Basket—Wire, for carrying hose, of ample capacity.
Valves—Two (2), heavy brass gate, and necessary brass piping.

Acid Receptacles—Four (4), noncorrosive metal.
Receptacle Holders—Two (2), in rear of engine, fitted with heavy brass covers.
Soda Bags—Two (2).
Tool Box—One (1), in front.
Hose Spanners—Two (2).
Wrenches—All necessary wrenches.
Lanterns—Two (2) Fire Department, with suitable holders.
Ax—One (1) fire department, with suitable holder.
Crowbar—One (1) fire department, with suitable holder.
Pressure Gauges—Two (2).
Gong—One (1), 10" automatic wheel striking, attached.
Painting—English vermilion, striped.
Striping—Gold leaf.
Lettering—To order.

Capacity—Two (2) 35-gallon cylinders.
Height—Over all, $54\frac{3}{4}$ inches.
Width—Over all, $65\frac{5}{8}$ inches.
Length—Over all, 8 feet $10\frac{1}{2}$ inches.
Gauge or Track—$58\frac{3}{8}$ inches.

Price, $1,000.00

CHEMICAL APPARATUS FOR MINE PROTECTION

It has been demonstrated that while much carbon dioxide is actually discharged into the fire, a considerable amount of the contents of the engine reaches the fire in liquid form. When it strikes the fire the heat of the fire converts the soda in solution into dioxide gas which, excluding the air, kills the fire. Further, when the fire is in the "breasts" or "headings," it may be fought there, and little or no attention need be given to the incipient fires that from time to time start in the "gob." Carbon dioxide is heavier than air, and as it accumulates in the vicinity of the fire it will spread out over the floor of the mine and smother any minor fires that may have started, so that the men will not be compelled to beat them out with cap and coat.

BADGER

Catalog No. A 1005

40-GALLON CHEMICAL ENGINE
HORIZONTAL TYPE

THIS engine is designed to be used either indoors or outdoors. Though strongly built, it is comparatively light, and its high wheels make it easy for two or three men to haul it, and it can be depended upon to perform effective general fire duty. It operates in the same manner as the engines heretofore mentioned, being only necessary to turn it over backwards and it is ready for service. The same high grade of excellence, both in material and labor, distinguishes this apparatus, and its finish is strictly in harmony with the general character of the equipment.

TANK: Capacity, 40 gallons. Made of seamless, high-pressure steel, drawn cold and pressed into shape; has double coating of lead, and tested to three times its required strength; all fittings on tank are of brass and highly polished. To operate, turn over backwards.

ACID BOTTLE: One, of chemically pure lead in brass case, with glass neck, lead stopper, and handle on top.

SODA BAG: One, of heavy canvas.

HOSE: 50 feet of ¾" 4-ply chemical hose all in one piece, no couplings except on ends.

NOZZLE: One, of brass, shut-off style.

FRAME: Of forged steel, with folding steel rest and handle bar on end.

DIMENSIONS: Outside width, 34", unless otherwise ordered; height, 55", weight, 700 lbs., loaded.

PAINTING: English Vermilion on wheels, frame tank, and gold striping.

This engine can be equipped with copper tank, roller-bearing axles and rubber tires at an additional cost.

The OBENCHAIN-BOYER CHEMICAL ENGINE

IN designing and constructing a Chemical Fire Engine, the fact which must be kept in mind is that it is a machine which is to be used under the most trying circumstances, when the operators will be working under a strain of excitement, when any hitch or delay will prove fatal to the purpose which the engine is designed to serve, that whenever it is called upon, it must, without fail, instantly and vigorously respond, and that, from the nature of the service, it will be subject to the severest kind of usage.

Single 45-Gallon Factory Engine No. 87

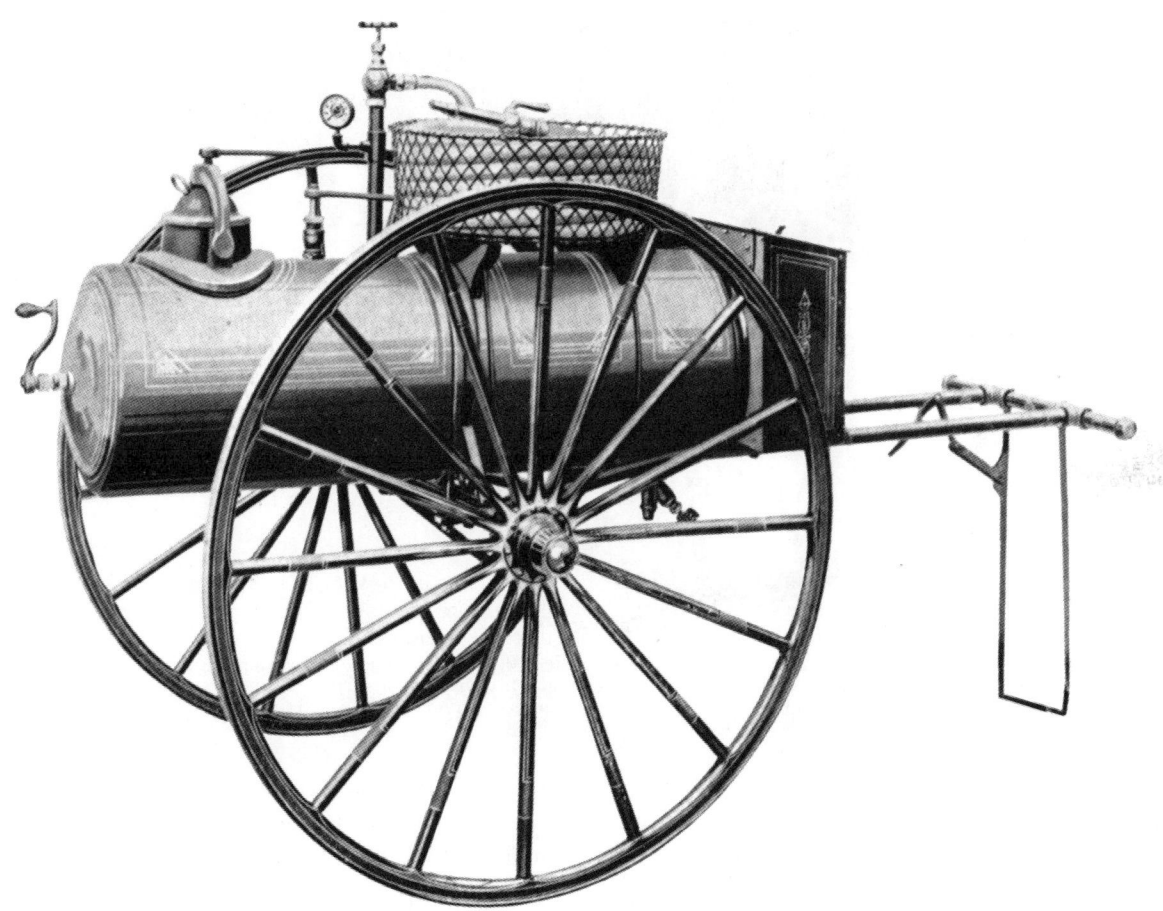

Catalog No. A 1010

DIMENSIONS—Height 4 feet 11 inches, length 7 feet 10 inches, width 4 feet 2 inches, tread 3 feet 4½ inches. Said engine to have one 45-gallon chemical tank, best flange steel, seams riveted and brazed, galvanized inside and out, tested to 400 pounds to the square inch, and equipped as follows:

WHEELS—Two, fire wheels, 48 inches in diameter, 1¼ inch tread, 1¼ inch steel axle.
AGITATOR—Flexible to dissolve soda, works on bottom of tank.
SAFETY CHARGING FLANGE—An absolute preventative to overcharging.
CHEMICAL BOX—One chemical box to carry two full charges.
HOSE BASKET—One metal hose basket to carry hose.
HOSE—50 feet of cotton web, rubber lined hose.
NOZZLE—One brass shut-off nozzle.
GAUGE—One pressure gauge.
CHARGING FUNNEL—One charging funnel.

This engine is also built with a narrow tread so as to pass through a 30 inch door.

The OBENCHAIN-BOYER CHEMICAL ENGINE

45-Gallon Portable Chemical Fire Engine No. 85

Catalog No. A 1011

DIMENSIONS—Height 4 feet 11 inches, length 10 feet 9 inches, width 4 feet, tread 3 feet 3 inches.
WHEELS—Two, best grade fire wheels, 52 inches high, 1⅜ inch spokes, rims 1¼ inches, 1⅜x5-16 inches round edge steel tires bolted and clipped to rims, points of hubs covered with polished brass hub caps.
AXLES—1¼ inch best steel, well trussed, hardened spindles.
SPRINGS—Two, 1½ inches by 36 inches, 4 leaves.
FRAME—Best steel channel side bars, hung on springs.
CHEMICAL TANK—One of 45 gallons capacity, best open hearth flange steel, rims riveted and brazed, tested to 400 pounds to the square inch.
AGITATOR—Flexible to dissolve soda, works on bottom of tank.
SAFETY CHARGING FLANGE—An absolute preventative to overcharging.
GAUGE—One on tank, registers up to 400 pounds.
CHEMICAL BOX—One, made of best sheet steel, for tools and charges.
HOSE—100 feet best cotton web, rubber lined cotton hose.
HOSE BASKET—One, made of woven wire, very substantial, plays out hose easily.
NOZZLE—One, shut-off pattern, heavy brass.
GONG—One, automatic, 12 inches in diameter, rings twice every revolution of wheel.
BUCKETS—Ten fire buckets, well painted and lettered.
LANTERN—One, solid brass, nickel plated and polished, with holder.
TOOLS—One fireman's axe with scabbard, crowbar and holder, monkey wrench and charging funnel.
REEL—One rope reel with 30 feet of ⅝ inch manilla pull rope attached.

The six points to be looked for in passing judgment on a Chemical Fire Engine are as follows:

Absolute sureness of action.
Speed and ease of operation.
Provision for dissolving the soda.
Provision for sealing the acid.
Security against overcharging.
Grade of material used, standard of workmanship, general appearance and finish.

With the above points in mind we invite you to the consideration of the Obenchain-Boyer Chemical Fire Engine.

The OBENCHAIN-BOYER CHEMICAL ENGINE

Double Tank 45-Gallon Portable Chemical Fire Engine No. 70

Catalog No. A 1012

WHEELS—Best grade, fire wheels 54 inches high, 1⅝ inch spokes, rims 1⅞ inches, 2x⅜ inch round edge steel tires bolted and clipped to rims, points of hubs covered with polished brass hub caps.
AXLES—1⅝ inch best steel, drop pattern, hardened spindles.
SPRINGS—1¾ inch x 40 inches, 8 leaves, ribbed and oil tempered.
FRAME—Best steel channel side bars, hung on springs.
CHEMICAL TANKS—Two, each of 45 gallons capacity, best open hearth flange steel, seams riveted and brazed, tested to 400 pounds to the square inch and so connected that one tank may be recharged while the other is being discharged.
AGITATOR—Flexible to dissolve soda, works on bottom of tank.
SAFETY CHARGING FLANGE—An absolute preventative to overcharging.
GAUGE—One on each tank, registers up to 400 pounds.
CHEMICAL BOX—Made of best steel, well riveted, to carry tools and charges.
HOSE—100 feet best 13-16 inch cotton web, rubber lined chemical hose. (Underwriters Specifications.)
HOSE BASKET—Made of woven wire, very substantial, plays out hose easily.
NOZZLE—Shut-off pattern, made of heavy brass.
GONG—Automatic, 12 inch diameter, rings twice every revolution of wheels.
BUCKETS—10 fire buckets, well painted and lettered if desired.
LANTERNS—Two, solid brass, polished and nickel plated, hung in suitable brackets.
AXE—One fireman's pickhead axe and holder.
REEL—One rope reel with 30 feet of ⅝ inch manilla pull rope attached.
TOOLS—One crowbar and holder, one monkey wrench and one charging funnel.

OPERATION OF ENGINE

When this type of engine is drawn to the scene of a fire there is but one thing necessary to set it in operation, that is, to pull, through a quarter turn, the operating lever on top of the tank. This lever actuates the bottle-breaking mechanism and within ten seconds from the time it is operated the pressure gauge will register from 140 to 180 pounds pressure.

BOTTLE BREAKING DEVICE

The device for breaking the bottle is so simple that it will, in all cases, positively effect its purpose. The bottle holder is so constructed that it retains all the broken glass and not a particle of it can fall into the tank.

No. 600
"Pirsch" Chemical Engine

SPECIFICATIONS

This engine is designed to meet all the requirements of a Volunteer Fire Department. Its high wheels and well-balanced construction make it easy running and well adapted for covering a large territory. Due to the simplicity of the construction of the tank, no experience is necessary to operate it; invert the tank, open valve and it is ready for service. The material and labor entering into its construction are of the best, the finish and painting are distinguished by the same grade of excellence, and are strictly in harmony with the general character of the apparatus. No modern feature is omitted and it is guaranteed to do effective and well sustained fire duty.

ENGINE: Has brass trunions on ends of tank resting on journal boxes on frame, tipping wheel for turning tank to operate it, brass shut-off valve to hose connection. All fittings on tank are of brass and highly polished.

TANK: Capacity 40, 50 or 60 gallons, made of seamless high-pressure steel, drawn cold and pressed into shape; has a double coating of lead and tested to three times its required strength.

PRESSURE GAUGE: 2½".

ACID BOTTLE: Two, of chemically pure lead, in brass case with glass neck, lead stopper, and handle on top.

SODA BAGS: One, of heavy canvas.

HOSE: 100 feet of ¾", 4-ply chemical hose.

NOZZLE: One, of brass, shut-off style.

HOSE BASKET: Made of iron with top rim made of tubing, so that hose will pull out freely. Basket is mounted on frame with iron standards.

HOLDERS FOR CHARGES: Two, one on each side of frame in rear.

AXE AND CROWBAR: One of each, held to side of frame in spring holders.

LANTERNS: Two, held in spring holders.

WHEELS: Sarven "A" grade, 56" high, with polished brass closed end hub caps.

TIRES: Round edge steel.

AXLES: Special Concord steel with solid collars.

SPRINGS: Semi-elliptic, best oil-tempered.

FRAME: Forged steel with steel tongue, handle bar, hand grabs and folding tongue rest.

GONG: 11" automatic.

ROPE REEL: With Flanges and ratchet, attached to front of frame, has 60 feet of manila rope.

TOOL BOX: Of suitable size; hung on front end of frame.

WRENCHES: All necessary wrenches and spanners are furnished.

DIMENSIONS: Tread, 48"; outside width, 58"; height, 57"; length, 10'.

PAINTING: English vermillion on wheels, frame and tank gold striping.

LETTERING: As ordered.

This engine can be equipped with copper tank, roller-bearing axle and rubber tires at additional cost.

Prices

40-gal	$450.00	50-gal	$475.00
60-gal			$500.00

Peter Pirsch & Sons Co.

No. 600 R. C.
"Pirsch" Chemical Engine

SPECIFICATIONS

Special attention is called to our Improved Caster Leg (Patented.) This leg is of trussed construction and can be folded up under the tongue, or allowed to remain in fixed position, being held by "U" shaped bracket and cotter pin. On the bottom of this leg rest is a universal caster, which allows the engine to be moved in a complete circle, forward and backward, without touching the leg support. This feature will be fully appreciated by those having had experience with the other style support, which has a tendency to bend or buckle and is usually replaced by a box, barrel, etc.

ENGINE: Has brass trunnions on ends of tanks, resting on journal boxes on frame, tipping wheel for turning crank to operate it, brass shut-off valve to hose connection. All fittings on tank are of brass and highly polished.

TANK: Capacity, 40, 50 or 60 gallons, made of seamless high pressure steel, drawn cold and pressed into shape; has a double coating of lead, and tested to three times its required strength.

PRESSURE GAUGE: 2½-inch brass case.

ACID BOTTLES: Two, of chemically pure lead, in brass case with glass neck, lead stopper, and handle on top.

SODA BAGS: One, of heavy canvas.

HOSE: 100 feet of ¾-inch 4-ply chemical hose.

NOZZLE: One, of brass, shut-off style.

HOSE REEL: Automatic hose reel, having brass stuffing box, and handle or crank for winding up hose. Hose so arranged that it is attached to reel at all times, ready for instant use, the principle being the same as that of the reels used on the large horse and motor apparatus.

HOLDERS FOR CHARGES: Two, one on each side of frame in front.

AXE AND CROWBAR: One on each, held on side of frame in spring holders.

LANTERNS: Two, held in spring holders.

WHEELS: Sarven "A" grade, 56 inches high, with polished brass closed end hub caps.

TIRES: Round edge steel.

AXLES: Special Concord steel with solid collars.

SPRINGS: Semi-elliptic, best oil tempered.

FRAME: Forged steel with steel tongue, handle bar, hand grabs and folding tongue rest, of trussed construction, having universal caster.

GONG: 11-inch automatic, nickel plated.

ROPE REEL: With flanges and ratchet, attached to front of frame; has 60 feet of manila rope.

TOOL BOX: Of metal and of suitable size; hung on front of frame.

WRENCHES: All necessary wrenches and spanners are furnished.

DIMENSIONS: Tread, 48 inches; outside width, 58 inches; height, 57 inches; length 10 feet.

PAINTING: English vermillion on wheels, frame and tank, gold striping, best wearing varnishes.

LETTERING: As ordered.

This engine can be equipped with copper tank, roller-bearing axle and rubber tires at additional cost.

Prices

40-gallon size	$465.00
50-gallon size	490.00
60-gallon size	515.00
Extra for roller bearings	30.00
Extra for rubber tires	40.00

No. 800
"Pirsch" Chemical Engine
Double Tank

SPECIFICATIONS

ENGINE: Has brass trunions on ends of tanks, resting on journal boxes on frame, tipping wheels for turning tanks to operate them, brass shut-off valve to hose connection. All fittings on tanks are of brass and highly polished.

NOTE: May be equipped with "Holloway" or "Pirsch" type tanks if preferred, at additional cost.

TANKS: Two, 30, 35 or 40 gallons, made of seamless high pressure steel, drawn cold and pressed into shape; have a double coating of lead and tested to three times their required strength.

PRESSURE GAUGES: Two, 2½".

ACID BOTTLES: Four, of chemically pure lead, in brass case with glass neck, lead stopper, and handle on top.

SODA BAGS: Two, of heavy canvas.

HOSE: 100 feet of ¾", 4-ply chemical hose.

NOZZLE: One, of brass, shut-off style.

HOSE BASKET: Made of iron with top rim made of tubing, so that hose will pull out freely. Basket is mounted on frame with iron standards. Equipped with hose reel if desired, at extra cost.

HOLDERS FOR CHARGES: Four, one on each side of frame in front and rear.

AXE AND CROWBAR: One of each, held to side of frame in spring holders.

LANTERNS: Two, Fire Dept. style, held in spring holders.

WHEELS: Sarven "A" grade, 4' 4" high, with polished brass closed end hub caps.

TIRES: Round edge steel.

AXLES: Special Concord steel with solid collars.

SPRINGS: Semi-elliptic, best oil tempered.

FRAME: Forged steel, with steel tongue, handle bar, hand grabs and folding tongue rest.

GONG: 11" automatic.

ROPE REEL: With flanges and ratchet, attached to front of frame, has 60 feet manila rope.

TOOL BOX: Of suitable size, on front end of frame.

WRENCHES: All necessary wrenches and spanners are furnished.

DIMENSIONS: Tread, 56"; outside width, 66"; height, 57"; length, 10'.

PAINTING: English vermillion on wheels, frame and tank, gold striping.

LETTERING: As ordered.

This engine can be equipped with copper tanks, roller-bearing axle and rubber tires at additional cost.

Price
Double 30........$725.00   Double 35......$750.00
Double 40..$800.00

PETER PIRSCH & SONS CO.

"Pirsch" Combination Hose, Chemical and Ladder Cart

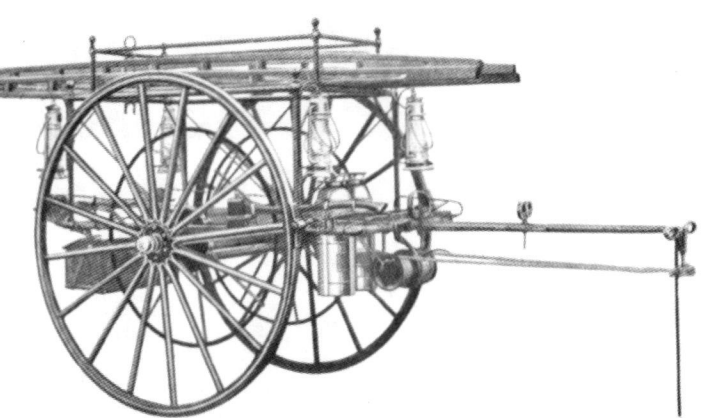

Description: This cart is designed to meet all the requirements of a **Village Fire Department**, and is the most efficient piece of hand apparatus built, combining hose, chemical and ladder; all necessary in fighting fire successfully.

The material and labor entering into its construction are of the best, the finish and painting being distinguished by the same grade of excellence and are strictly in harmony with the general character of the apparatus.

FRAME: Equipped with tongue, handle bar, hand grabs and folding tongue rest. Tongue equipped with two play pipe holders. Roller in rear.

WHEELS: Heavy "Archibald" Artillery Wheels. Polished brass hub caps. 5 ft. 6 in. or 6 ft. high.

AXLES: Special Concord steel with solid collars or roller bearing.

TIRES: Round edge steel.

BRAKES: Shoe brake for wheels and friction band brake for Hose reel.

HOSE REEL: Capacity, 750 ft. 2½ in. Heavy Regulation Fire Department Hose. Equipped with chain winding attachment and winch handle for reeling on the hose.

ROPE REEL: With flanges and ratchet attached to front of frame, equipped with 50 ft. Manila rope.

CHEMICALS: Two three-gallon "Underwriters" Fire Extinguishers carried in suitable holders.

LADDERS: One 25 ft. extension ladder in three sections. Securely carried on top of cart.

TOOL BOX: One wooden tool box in rear. Compartment for carrying extinguishers, charges, wrenches, etc.

LANTERNS: Four Fire Department lanterns in suitable holders.

AXE: One in suitable holder.

CROW BAR: One in suitable holder.

PAINT: English vermillion; gold stripe.

DIMENSIONS: Length, 12 ft.; width 5 ft. 5 in.; height 6 ft.; weight 875 pounds. Crated 1000 pounds. Price **$450.00**

VI

HORSE DRAWN CHEMICAL FIRE ENGINES

Horse drawn chemical engines did not evolve from hand drawn engines; the horse drawn engines were manufactured right from the start by Babcock and Holloway and later by many others. The following pages show drawings and photographs of horse drawn chemical engines from some of the earliest models right up to the motorized era.

City and Village Chemical Fire Engines.
"CHAMPION."—"HOLLOWAY."—"BABCOCK."

Plate 314.

5,000 IN SERVICE IN SUCH CITIES AS

New York, Lowell,
Chicago, Brockton,
Brooklyn, Haverhill,
Philadelphia, Indianapolis,
Baltimore, Worcester,
Washington, Holyoke,
St. Louis, Lynn,
Boston, Des Moines,
San Francisco, Dubuque,
Cincinnati, Fort Worth,
Cleveland, Jacksonville,
Buffalo, Los Angeles,
Syracuse, Memphis,
Oswego, San Diego,
New Orleans, Sioux City,
Columbus, Wilmington, N. C.
Pittsburg, Norfolk,
Detroit, Wheeling,
Milwaukee, Rochester,
Minneapolis, Kansas City,
Jersey City, Denver,
Omaha, Charleston,
St. Paul, Liverpool, Eng.
Louisville, Etc., etc., etc.

Plate 316.

CHAMPION BABCOCK.
Combination Chemical Engine and Hose Wagon.

The evolution of the old time hose reel to the modern Combination Chemical Engine and Hose Wagon marks an interesting and notable advance in the manufacture of fire department appliances.

The very general adoption of our Combination Wagons by the leading fire departments in the United States and abroad, and their remarkable achievements in service, have demonstrated that this type of fire apparatus is a most valuable acquisition to any department.

The wagon has a hose capacity of from 800 to 1000 feet of cotton rubber-lined hose, and carries a varied and complete assortment of tools and patented specialties.

A powerful chemical engine attachment is conveniently located under the driver's seat, where it is always ready for immediate service. This combination is virtually two pieces of apparatus in one, so that in addition to an increase in efficiency, there is a saving in both initial cost, running expenses and house room.

"Holloway"
Double Cylinder Chemical Fire Engine.

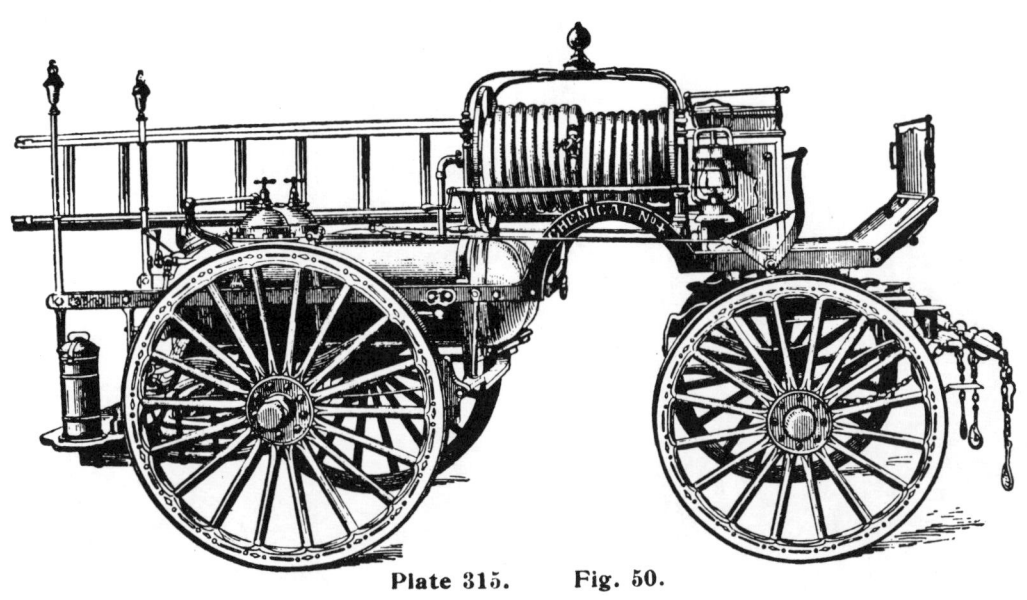

Plate 315.   Fig. 50.

HOLLOWAY DOUBLE TANK CHEMICAL ENGINES.
Horizontal Tanks.

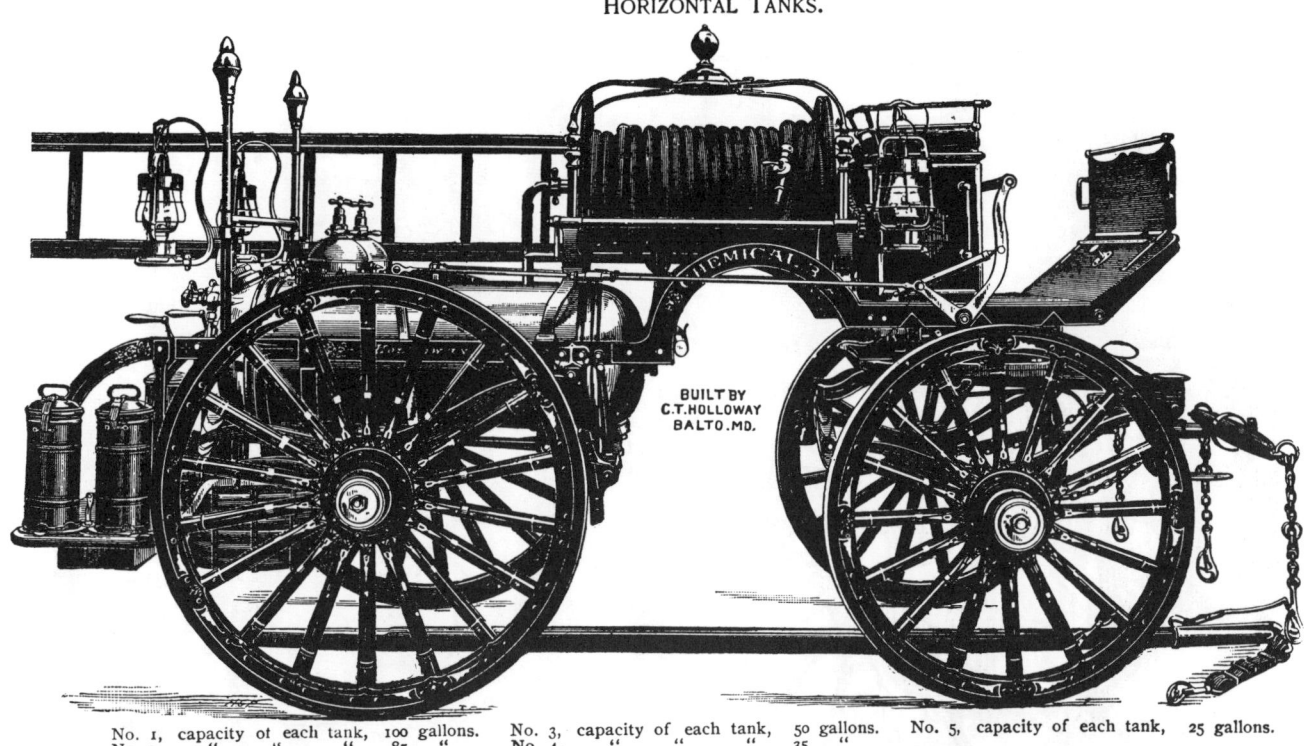

No. 1, capacity of each tank, 100 gallons.   No. 3, capacity of each tank, 50 gallons.   No. 5, capacity of each tank, 25 gallons.
No. 2,   "   "   "   "   85   "   No. 4,   "   "   "   "   35   "

No. 4 Chemical Engine for Cities and Towns.

"Holloway" Combination Chemical Engine and Hose Wagon.

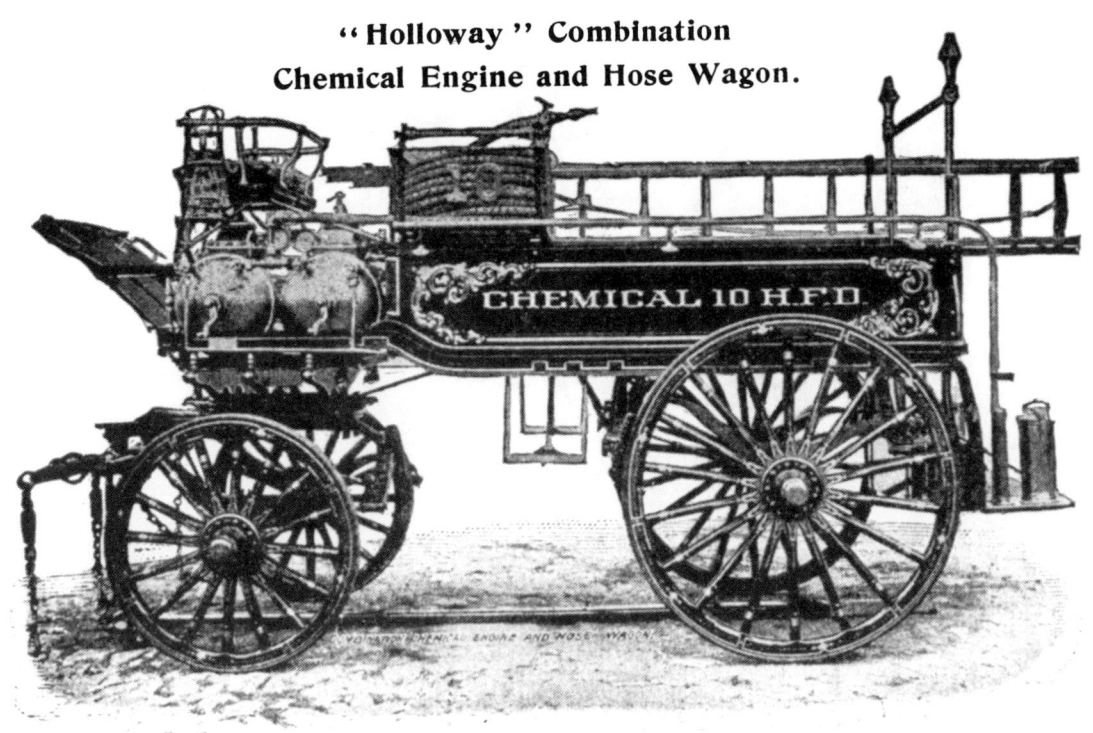

Holloway Double Tank Chemical Engines.
HORIZONTAL TANKS.

| No 1, CAPACITY OF EACH TANK, 100 GALLONS. |
| " 2,   "   "   "   "   85   " |
| " 3,   "   "   "   "   50   " |
| " 4,   "   "   "   "   35   " |
| " 5,   "   "   "   "   25   " |

HOLLOWAY

Single Tank Chemical Engine, with Hook and Ladder Attachments.

Combination Hook and Ladder Truck with Chemical Engine Attachment.

COMBINATION "CHAMPION" CHEMICAL ENGINE AND HOSE OR PATROL WAGON.
(Style 203.)

CHAMPION

COMBINATION ALL-STEEL "CHAMPION" CHEMICAL ENGINE AND HOSE WAGON.

(Style 201.)

The wagon body, sides and front made of angle steel. The bed of perforated steel is so constructed as to allow air to pass through from all directions for drying the hose and preventing mildew. The chemical engine attachment makes this the most perfect piece of fire apparatus in existence.

WOODEN BED COMBINATION "CHAMPION" CHEMICAL ENGINE AND HOSE WAGON.
(Style 200.)

WOODEN BED COMBINATION "CHAMPION" CHEMICAL ENGINE AND HOSE WAGON. (Rear View.)
(Style 200.)

SINGLE CYLINDER ONE-HORSE TWO-WHEEL "CHAMPION" CHEMICAL ENGINE.
(Style 74.)

DOUBLE CYLINDER FOUR-WHEEL "CHAMPION" CHEMICAL ENGINE. (Style 58.)
(The adopted Chemical Engine of the "World's Columbian Exposition," Chicago, 1893.)

DOUBLE CYLINDER FOUR-WHEEL (SPECIAL) "CHAMPION" CHEMICAL ENGINE.
(Wire Basket for Hose.) (Style 75.)

DOUBLE CYLINDER (HORIZONTAL) "BABCOCK" CHEMICAL ENGINE.
(Style 13.)

DOUBLE CYLINDER COMBINATION "CHAMPION" ENGINE AND HOSE WAGON.
(With Hook and Ladder Attachment.)   (Style 70.)

DOUBLE UPRIGHT CYLINDER "BABCOCK" CHEMICAL FIRE ENGINE.
(Style 6.)

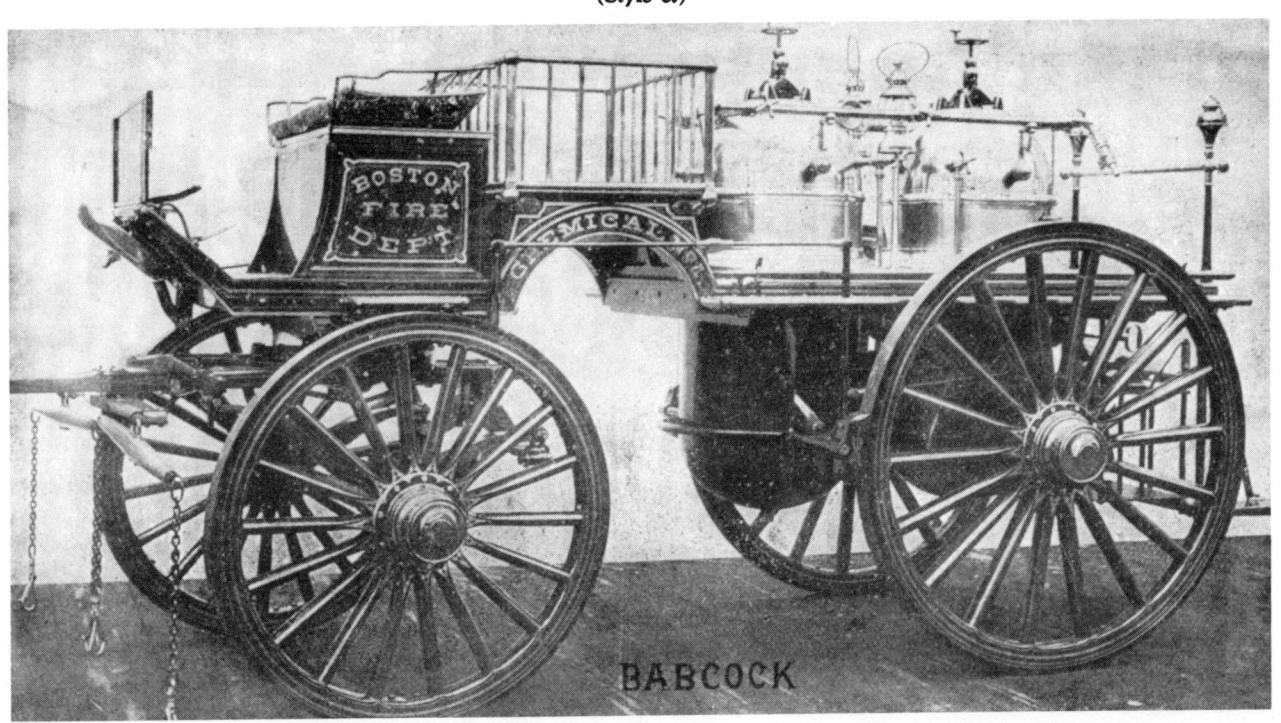

COMBINATION "CHAMPION" CHEMICAL ENGINE AND HOSE CARRIAGE. (Style 202.)

COMBINATION "BABCOCK" CHEMICAL ENGINE AND HOSE CARRIAGE.
(Style 206.)

"CHAMPION-BABCOCK" COMPOUND CHEMICAL FIRE ENGINE (Special).

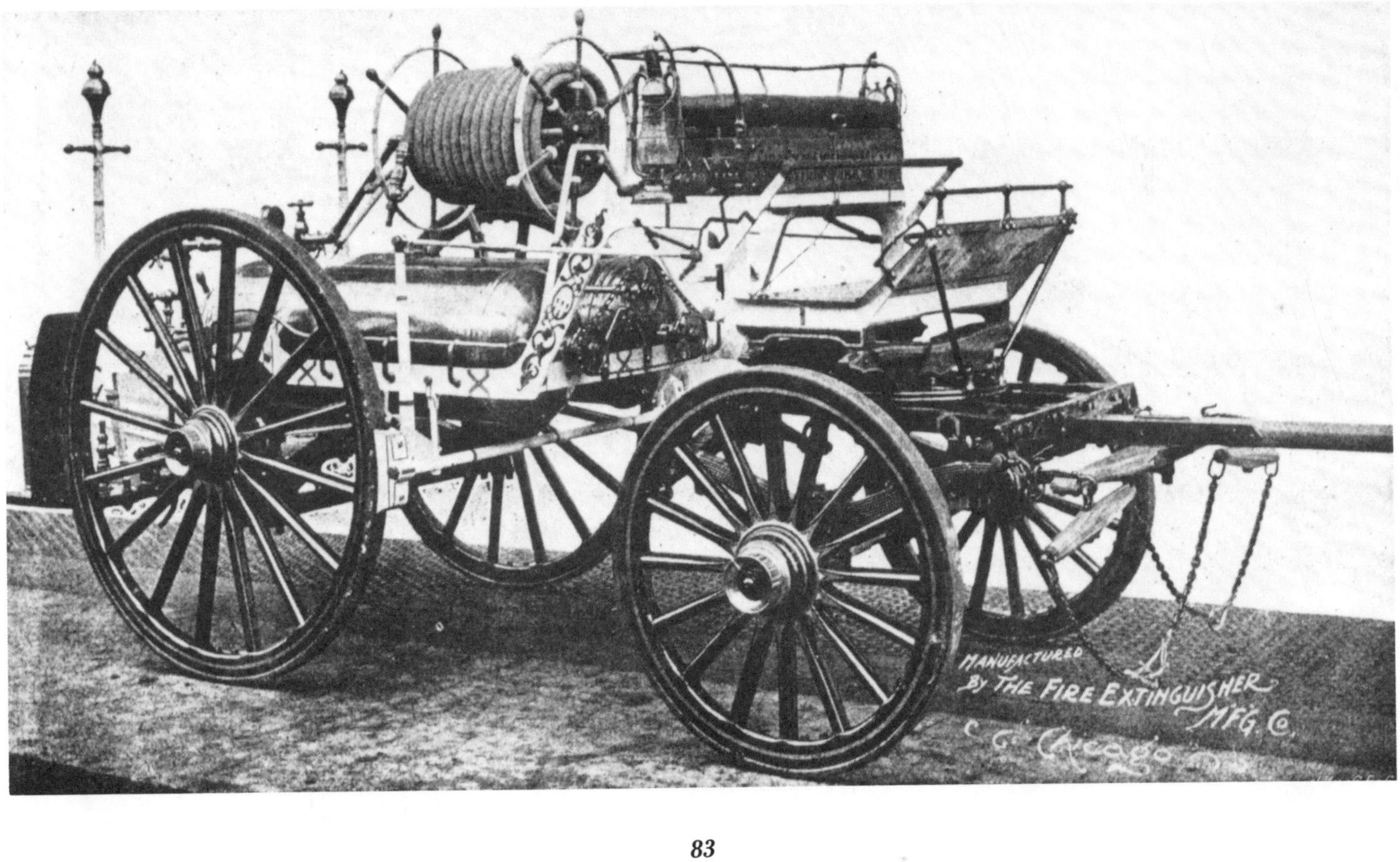

COMBINATION "BABCOCK" ENGINE AND HOSE OR PATROL WAGON.
(Style 205.)

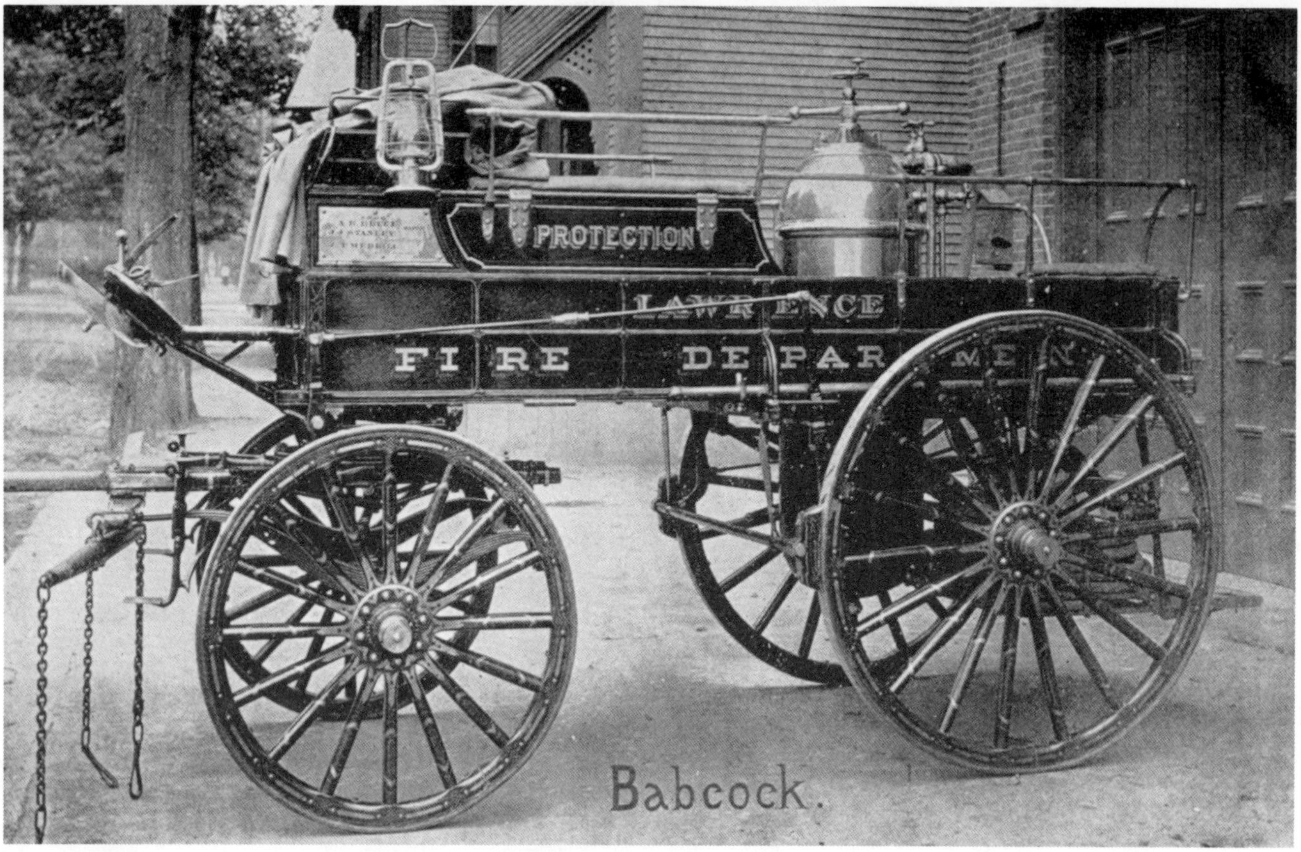

COMBINATION HOOK AND LADDER TRUCK, CHEMICAL ENGINE AND HOSE WAGON.
(Style 107.)

J.E. McWILLIAM & SON
HUBBARDSTON, MASSACHUSETTS

Fig. 12.—Two Horse Chemical Truck of 220 gallon capacity.

Since placing our machines on the market, their great success as a fire annihilator has brought them renown from all sides, from resident, factory, farm and town. And while our sole idea last year was to put before the public our small portable and large stationary automatic fire extinguishing tanks, we have been forced to add to our list two extinguishing trucks, a large one of 220 gallon and a small one of 25 gallon capacity.

We have been forced to add these because of the constant demand for such apparatus, and can now supply villages, towns and cities with the best chemical fire appliances for the least money in the market. On page 30, Fig. 12, appears our larger two horse hitch, built to stand rough roads and hard usage. All woodwork is of seasoned white oak and maple, the wheels are made extra strong and heavy. The tank is of heavy steel plate, boiler construction, placed horizontally to give it a low center of gravity besides a neat and trim appearance. The dome at the top gives the same ease and convenience in charging as our stationary tanks. It also contains the acid bottle which is held in place by a yoke shaped basket, one end protruding outside through packing box and having lever attached thereon. To operate it is only necessary to pull the lever a quarter turn which tips bottle inside, spilling the acid down through the bicarbonate solution, causing an instant pressure of 200 lbs.

per square inch and throwing a stream 70 ft. high. The tank holds 220 gallons, will throw a quarter inch stream of chemicals 45 minutes. With our patent shut-off nozzle it can be turned off or on at will and used to the best advantage. It quite often happens that a small fire does not require much to extinguish it, and in view of this case we furnish with each large truck two five gallon chemical tanks, one on either side of the driver's seat. This saves the necessity of discharging the large tank on a small fire. The engine is handsomely painted in dark green with chrome yellow running gear, striped in red and black. A heavy brass railing (see cut) surrounds the tank, contrasting strongly with the dark colors and making a handsome apparatus for any town or city. The hose box is carried behind where it is convenient, with the discharge pipe connected at all times to the hose. A step on the rear affords ample room for two or three passengers.

If your town is contemplating putting in chemical engines write and let us tell you more about it, as the price is extremely low.

To charge the above, dissolve 96 lbs. of bicarbonate of soda in water enough to fill within 3 inches of the bottle bar, fill bottle with sulphuric acid, fill grove with heavy screw-on-cover, and all is ready.

VII

MOTORIZED CHEMICAL FIRE ENGINES

As internal combustion engines began to replace horses in America's firehouses during the second decade of the twentieth century, chemical engines did not lose favor. Instead, they proliferated all the more, the only difference being the means of getting them to the fire.

Although gasoline engines were now moving the chemical engines to the fire, it was still the time-honored chemical formula $H_2SO_4 + NaHCO_3 \longrightarrow CO_2$ that shot the soda water from the chemical tanks onto the fire, and it was still widely believed that the "chemicals" were 30 to 40 times more effective than plain water.

Chemical apparatus was mounted not only on Ford Model T's, but on just about every type of motorized vehicle that was for sale, and there were soon dozens of manufacturers. American LaFrance and Seagrave built their own. American LaFrance also sold do-it-yourself kits of chemical apparatus for firemen to build their own chemical engines on Model T chassis.

In addition to straight "chemical cars" there were soon chemical and hose "double combinations" and chemical, hose and volume pump "triple combinations". Of course the volume pump, whether front mounted or midship, and whether rotary, piston, or centrifugal, was powered by the same gasoline engine that moved the vehicle.

It was a whole new ballgame to America's firehouses, but chemical engines largely continued to be that important one-third of most triple combinations on through the 1920's and, on a few, even into the early 1930's. The following pages depict some of the early motorized chemical cars and chemical double and triple combinations.

The transition from horse drawn to motorized chemical engines had two short-lived aberrations, the first when, in 1904, American LaFrance delivered a *steam-powered* combination chemical and hose car to Niagara Fire Co. No. 1 in New London, Conn. This may well have been the only chemical rig with a smokestack.

A "LA FRANCE" CHEMICAL FIRE ENGINE PROPELLED BY INDIVIDUAL STEAM ENGINES GEARED TO THE REAR WHEELS. THIS ENGINE BELONGS TO THE FIRE DEPARTMENT OF NEW LONDON, CONN.

The second brief aberration in the horse to motorized transition occurred in 1910 when Ahrens-Fox produced a "Continental" Fire Chief's auto which carried chemical equipment which was *battery-powered*. Electricity was the energy source that got the rig to the fire. It is ironic that just three years later the invention of the booster system by Mr. Fox of Ahrens-Fox led to the ultimate demise of chemical engines.

Fire service historians generally concede that the first motorized fire engine was a Waterous delivered to the Wayne, Pa. fire company in 1906. But perhaps it was a tie, as the Knox Automobile Co. of Springfield, Mass. exhibited and sold the first motorized chemical engines also in 1906. These Knox passenger cars converted by Knox into chemical cars carried two 35 gallon chemical tanks, and could attain a top speed of 40 miles an hour.

Some of the first manufacturers of motorized chemical apparatus during the first decade of the motorized era are as follows:

DATE	MANUFACTURED	TYPE
1906	Knox Automobile Co. Springfield, Mass.	straight chemical with two 35 gallon tanks
1907	Seagrave	straight chemical delivered to Vancouver, B.C., Canada
1907	American LaFrance	straight chemical built on a Packhard chassis delivered to Boston Fire Dept. Top speed 35 MPH.
1907	Al. C. Webb Joplin, Mo.	straight chemical car on Buick chassis. One 60 gallon tank. Delivered to Joplin Fire Dept.
1908	Locomobile	straight chemical. One 50 gal. tank. Delivered to Bridgeport, Conn. Fire Department
1909	Tea Tray Co. Newark, N.J.	first ever triple combination American Mo!rs Auto Chassis. 50 gal. chemical tank, hose bed, 400 GPM Gould pump. Delivered to Monhagen Hose Co., Middletown, N.Y.
1909	Pope Manufacturing Co. Hartford, Conn.	combination chemical & hose to Hartford, Conn. Fire Department

DATE	MANUFACTURER	TYPE
1909	Auto Car Co.	two tank chemical to E.H. Stokes Fire Co., Ocean Grove, N.J. Chemical tanks were manufactured by Holloway
1910	Brockway Motor Truck Cortland, N.Y.	chemical & hose combination 40 H.P. Priced at $2750.00
1910	James Boyd and Brothers Philadelphia, Pa.	chemical & hose combination for Pioneer Fire Co., No. 1, Jenkintown, Pa. Two chemical tanks.
1911	Peter Pirsch & Sons Co., Kensoha, Wisc.	chemical & hose combination on a Rambler chassis for Aurora, Ill.
1912	Pope-Hartford	straight chemical. Two 60 gallon tanks. For Lynn, Mass. Fire Department
1912	Harder Auto Truck Co., Chicago, Ill.	chemical & hose combination for Chicago Fire Dept.
1913	U.S. Fire Apparatus Co., Wilmington, Del.	two tank chemical for Water Witch Fire Co. No. 5 of Wilmington, Del.
1914	Maxim Motor Co. Middleboro, Mass.	chemical & hose combination on an E.R. Thomas automobile chassis for Ansonia, Conn. Fire Department
1914	Davis Sewing Machine Co., Datyon, Ohio	Dayton "Tricar" motorcycle with chemical tank
1914	Anderson Coupling and Fire Supply Co., Kansas City, Kans.	single tank chemical & hose combination
1916	Mack	chemical & hose combination to Lakewood, N.J. Fire Department

American-LaFrance Fire Fighting Equipment

MOTOR FIRE APPARATUS FOR SMALL CITIES AND TOWNS

Plate 839
American-LaFrance Chemical on Standard Ford Chassis

SPECIFICATIONS

Type "A"
Double-Tank Chemical Car

Read carefully the following specifications of the double-tank chemical. It's a car remarkable for its simplicity, reliability, ease of operation and quickness of action. The chemical stream can be thrown for a distance of about 60 to 75 feet. Since the nozzle is not left continuously open like that of a water hose—"time out" being called constantly—this means that each tank is capable of an immense amount of fire fighting.

CHASSIS—Ford Motor Company's standard chassis, 4-cylinder, 3¾" x 4", 20 H. P.

TANKS—Two (2), 25-gallon capacity each. Champion style. Seamless drawn steel.

HOSE—150 ft. ¾" special four-ply rubber chemical hose with heavy brass couplings attached.

NOZZLE—One (1) brass shut-off nozzle, eccentric with two (2) tips.

BASKET—One (1) wire hose basket.

ACID RECEPTACLES—Three (3) noncorrosive metal.

ACID RECEPTACLE HOLDER—One (1) brass, with cover, mounted on running-board.

SODA CANISTER—One (1) painted, with cover, mounted on running-board.

LANTERNS—Two (2) fire-department style, with suitable holders.

AX—One (1) fire-department style, with suitable holder.

CROWBAR—One (1), fire-department style, with suitable holder.

EXTINGUISHER—One (1), fire-department style, finished in polished copper.

EXTINGUISHER HOLDER—One (1), steel, mounted on running-board.

BELL—One (1), Locomotive type, mounted.

LADDER—One (1) 16-ft. solid side extension, finished in natural wood, complete with suitable holders.

HOSE SPANNERS—Two (2) chemical.

WRENCHES—All necessary wrenches.

SODA BAG—One (1).

PRESSURE GAUGES—Two (2).

PIPING—Brass. By-pass system.

LIGHTING SYSTEM—Two electric headlights, two oil side lights, one oil tail light.

TOOL BOX—One (1) mounted on running-board.

TIRE-PUMP—One (1).

JACK—One (1).

TOOLS—Complete set motor and chassis.

PAINTING—Entire car [fire-department] red, striped.

American-LaFrance Fire Fighting Equipment

Plate 2839
American-LaFrance Chemical Car on Ford One-Ton Chassis

Type "D"
Double Tank Chemical Car

This apparatus with its increased carrying and chemical capacity is proving to be an ideal piece of equipment for the small town and rural community. In farming districts this car has often paid for itself at a single fire in a grain field or timber tract.

SPECIFICATIONS

CHASSIS—Ford Model "T" one-ton chassis.

TANKS—Two (2) 35-gallon, seamless, cold drawn steel. Champion style.

CHEMICAL PIPING—Heavy brass; by-pass system with five valves. Two 350-pound pressure gauges.

HOSE—200 feet of ¾-inch special chemical hose with heavy brass couplings attached.

NOZZLE—One (1) brass shut-off nozzle with two (2) tips.

HOSE BASKET—Wire, with capacity for 250 feet of ¾-inch hose.

ACID RECEPTACLES—Three, (3) non-corrosive metal.

ACID RECEPTACLE HOLDERS—Two (2) brass, mounted on running board.

SODA CANISTERS—Two (2) heavy sheet steel, mounted on running board.

SODA BAGS—Two (2) heavy duck.

LANTERNS—Two (2) fire-department style.

AX—One (1) fire-department style.

CROWBAR—One (1) fire-department style.

PIKE POLE—One (1).

EXTINGUISHERS—Two (2) No. 5 Babcock.

EXTINGUISHER HOLDER—Two (2) steel, mounted on running-board.

BELL—One (1) locomotive type, mounted.

LIGHTING SYSTEM—Two electric head lights, two oil side lights, one oil tail light.

LADDERS—One 20-ft. extension ladder. One 12-ft. roof ladder.

TOOL BOX—One (1) mounted on running board.

TOOLS—Complete set for tires, motor and equipment.

PAINTING—Entire apparatus fire-department red, striped.

American-LaFrance Fire Fighting Equipment

ILLUSTRATION OF OUR STANDARD CHAMPION EQUIPMENT

FOR MOUNTING ON MOTOR CHASSIS

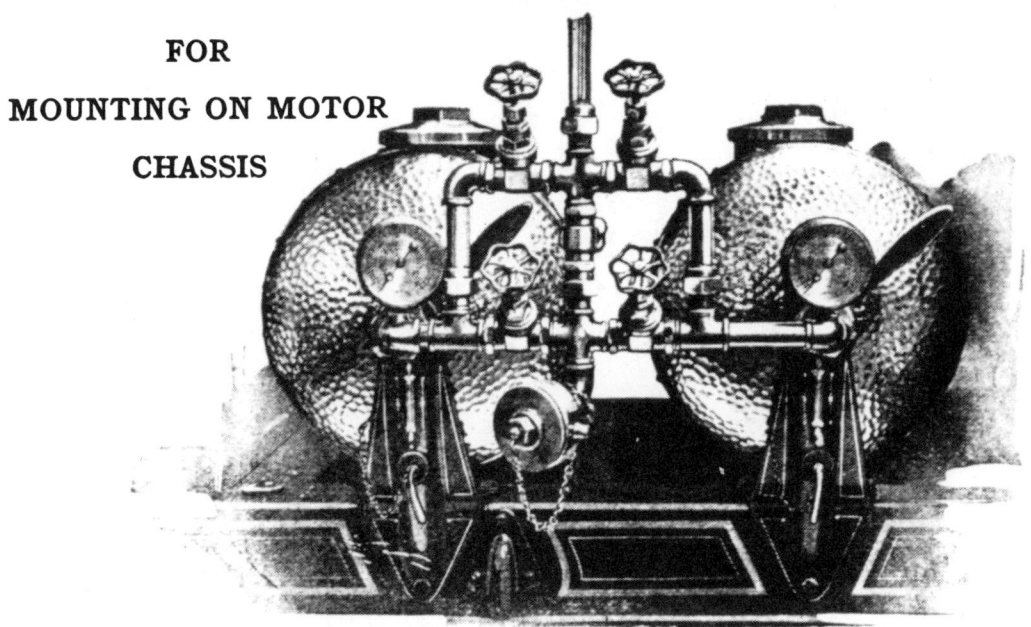

Plate 841

On our double tank equipment the piping is arranged so that one tank can be discharged while the other is being filled; also arranged with 2½-inch water filling connection. This connection is threaded to connect to city fire hose.

Plate 842 is our Standard Champion Single Tank equipment, with piping, filling connection, etc.

NOTE—Supporting brackets are not furnished with tanks since the latter are standard for our own apparatus and will not fit any other frame. Tanks are, however, fitted up completely with trunnions so that the customer can make up and attach his own supporting brackets.

Plate 842

American-LaFrance Fire Fighting Equipment

Plate 2840
American-LaFrance Combination Car on Ford One-Ton Chassis

Type "C"
Single Tank Combination Chemical and Hose Car

This American-LaFrance combination chemical and hose car is almost a complete Fire Department in itself. It has a generous hose-carrying capacity, and with the addition of the forty-gallon chemical tank, firemen are prepared for the big or little fire.

SPECIFICATIONS

CHASSIS—Ford Model "T," one-ton chassis.

HOSE BODY—Capacity, 1000 feet of 2½-inch double-jacketed fire hose.

CHEMICAL TANK—One (1) 40-gallon capacity, seamless cold drawn steel, Champion style.

CHEMICAL PIPING—Heavy brass, by-pass system. One 350-pound gauge.

CHEMICAL HOSE—200 feet of ¾-inch special chemical hose with heavy brass couplings attached.

NOZZLE—One (1) brass shut-off nozzle with two tips.

HOSE BASKET—Steel, with a capacity of 250 feet of ¾-inch hose.

ACID RECEPTACLES—Two (2) noncorrosive metal.

ACID RECEPTACLE HOLDER—One (1) brass, mounted on running-board.

SODA CANISTER—One (1) heavy steel, mounted on running-board.

SODA BAG—One (1) heavy duck.

LANTERNS—Two (2) Fire-Department Style.

AX—One (1) Fire-Department Style.

CROWBAR—One (1) Fire-Department Style.

PIKE POLE—One (1).

EXTINGUISHERS—Two (2), No. 5 Babcock.

EXTINGUISHER HOLDERS—Two (2) steel, mounted on running-board.

BELL—One (1) locomotive style, mounted.

LIGHTING SYSTEM—Two (2) electric head lights, two oil side lights, one oil tail light.

LADDERS—One 20-ft. extension ladder. One 12-ft. roof ladder.

TOOL BOX—One (1) mounted on running-board.

TOOLS—Complete set for tires, motor and equipment.

PAINTING—Entire apparatus fire department red, striped.

Plate 2840-E
American-LaFrance Combination Car on Ford One-Ton Chassis

Type "E"
Double Tank Combination Chemical and Hose Car

A large hose body and two chemical tanks provide this car with full fire-fighting equipment. The double tank arrangement, with by-pass system, permits using a continuous chemical stream if it is necessary.

SPECIFICATIONS

Chassis—Ford Model "T," on one-ton chassis.

Hose Body—Capacity, 1000 feet of 2½-inch, double-jacketed fire hose.

Chemical Tanks—Two (2), 25-gallon, seamless, cold drawn steel, Champion style.

Chemical Piping—Heavy brass, by-pass system with five valves. Two 350-pound-pressure gauges.

Chemical Hose—200 feet of ¾-inch, special chemical hose with heavy brass couplings attached.

Nozzle—One (1), brass shut-off with two tips.

Hose Basket—Steel, with a capacity of 250 feet of ¾-inch hose.

Acid Receptacles—Two (2), noncorrosive metal.

Acid Receptacle Holders—Two (2), brass, mounted on running-board.

Soda Canisters—Two (2), heavy steel, mounted on running-board.

Soda Bags—Two (2), heavy duck.

Lanterns—Two (2), Fire Department Style.

Ax—One (1), Fire Department Style.

Crowbar—One (1), Fire Department Style.

Pike Pole—One (1).

Extinguishers—Two (2), No. 5 Babcock.

Extinguisher Holders—Two (2), steel, mounted on running-board.

Bell—One (1), locomotive style, mounted.

Lighting System—Two (2) electric head-lights, two oil side-lights, one oil tail-light.

Ladders—One 20-foot extension ladder. One 12-foot roof ladder.

Tool Box—One (1), mounted on running-board.

Tools—Complete set for tires, motor and equipment.

Painting—Entire apparatus Fire-department red, striped.

American-LaFrance Fire Fighting Equipment

The Standard Chemical Tank Equipment for Combination Chemical Engine and Hose Motor Car consists of the following:

Chemical Tank, capacity 40 gallons when single tank is ordered; 35 gallons each when double tank is ordered.

Hose basket or reel.

200 ft. ¾-inch chemical hose.

1 shut-off nozzle with two tips.

2 hose spanners for chemical hose.

1 heavy duck soda bag.

1 extra acid receptacle with each tank.

1 brass acid receptacle holder for holding extra acid receptacle on step.

Pressure gauge, piping, and 2½-inch water-filling connection.

All necessary wrenches.

Variations in regard to equipment and accessories can be made to suit the customer. Always state exact requirements when asking for prices. We manufacture chemical tanks in all sizes from 25 gallons capacity up, both in single and double tank arrangements.

Prices quoted on request.

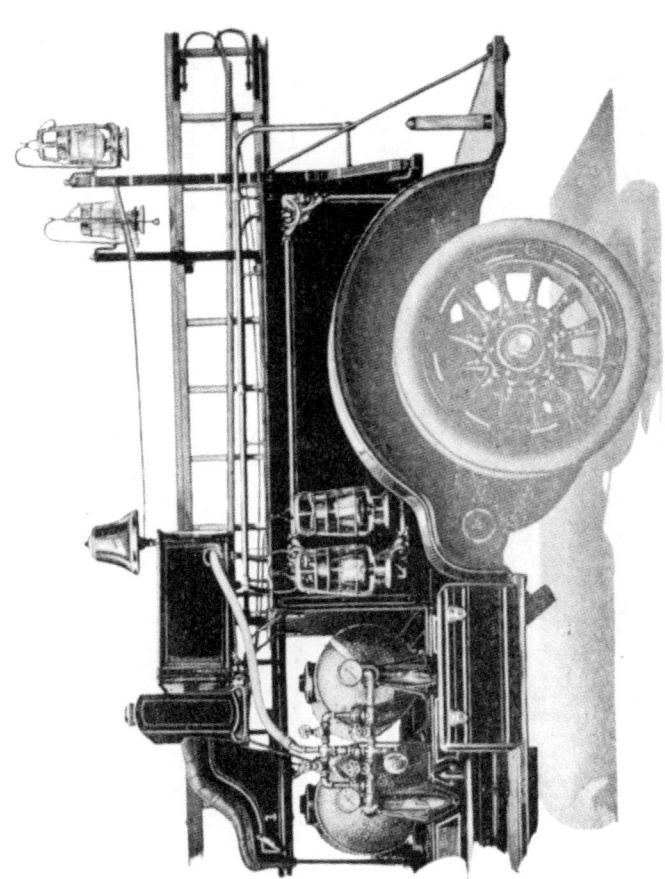

Plate 843

This illustration shows our method of mounting equipment. We are in position to furnish you with any or all of the items shown. Also note in the above illustration our Standard Champion Tanks (35-gallon capacity) with piping, 2½-inch filling connections, etc. The bodies are made of steel, and can be furnished with 1000 or 1200 ft. of 2½-inch fire hose capacity, the standard practice being 1000 feet when two tanks are mounted, and 1200 feet when single tank is mounted.

American-LaFrance Fire Fighting Equipment

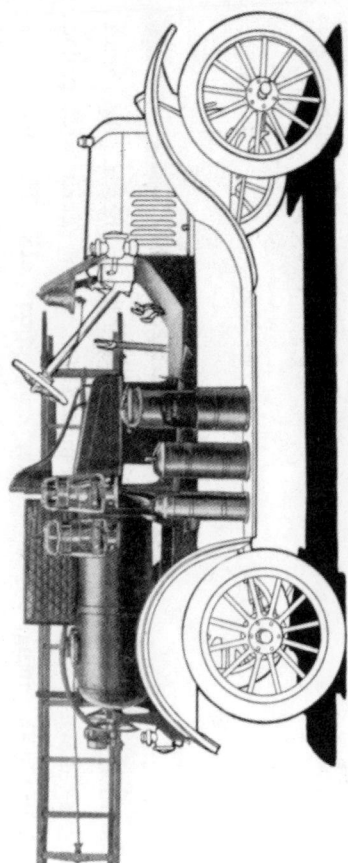

Plate 2840

American-LaFrance Combination Car Equipment Only

Type "C"
Single Tank Combination Chemical and Hose Car

This equipment is designed for mounting on the one-ton Ford chassis and contains all essential equipment for practical fire fighting. This equipment is furnished completely assembled ready to be quickly attached to the truck frame. All American-LaFrance equipment is ready for instant service, and embodies the completeness and careful workmanship that comes from successful experience in building fire fighting apparatus.

Equipment for the one-ton Ford Chassis can also be furnished for the same types illustrated on pages 137 and 139.

HOSE BODY—Capacity, 1000 feet of 2½-inch double-jacketed fire hose.

CHEMICAL TANK—One (1) 40-gallon capacity, seamless cold drawn steel, Champion style.

CHEMICAL PIPING—Heavy brass, by-pass system. One 350-pound gauge.

CHEMICAL HOSE—200 feet of ¾-inch special chemical hose with heavy brass couplings attached.

NOZZLE—One (1) brass shut-off nozzle with two tips.

HOSE BASKET—Steel, with a capacity of 250 feet of ¾-inch hose.

ACID RECEPTACLES—Two (2) noncorrosive metal.

ACID RECEPTACLE HOLDER—One (1), brass.

SODA CANISTER—One (1) heavy steel.

SODA BAG—One (1) heavy duck.

LANTERNS—Two (2), Fire-Department Style.

AX—One (1), Fire-Department Style.

CROWBAR—One (1) Fire-Department Style.

PIKE POLE—One (1).

EXTINGUISHERS—Two (2), No. 5 Babcock.

EXTINGUISHER HOLDERS—Two (2) steel.

BELL—One (1), locomotive style.

LADDERS—One 20-ft. extension ladder. One 12-ft. roof ladder.

TOOL BOX—One (1) mounted on running board.

PAINTING—Fire department red, striped.

Plate 839

American-LaFrance Chemical Equipment Only

Type "A"
Double-Tank Chemical Car

This equipment is for mounting on the standard Ford touring chassis, and is remarkable for its completeness and durable construction. The equipment can be easily mounted on the chassis frame by any experienced workman. The body sills are drilled, ready to be bolted quickly to the frame. A well-built low priced fire fighting outfit for the small town or large manufacturing plant.

SPECIFICATIONS

TANKS—Two (2), 25-gallon capacity each. Champion style. Seamless drawn steel.

HOSE—150 ft. ¾" special four-ply rubber chemical hose with heavy brass couplings attached.

NOZZLE—One (1) brass shut-off nozzle, eccentric with two (2) tips.

BASKET—One (1) wire hose basket.

ACID RECEPTACLES—Three (3) noncorrosive metal.

ACID RECEPTACLE HOLDER—One (1) brass, with cover, to be mounted on running-board.

SODA CANISTER—One (1) painted, with cover, to be mounted on running-board.

LANTERNS—Two (2) fire-department style, with suitable holders.

AX—One (1) fire-department style, with suitable holder.

CROWBAR—One (1), fire-department style, with suitable holder.

EXTINGUISHER—One (1), fire-department style, finished in polished copper.

EXTINGUISHER HOLDER—One (1) steel

BELL—One (1) Locomotive type.

LADDER—One (1) 16-ft. solid side extension, finished in natural wood, complete with suitable holders.

HOSE SPANNERS—Two (2) chemical.

WRENCHES—All necessary wrenches.

SODA BAG—One (1).

TOOL BOX—One (1) mounted on running-board.

PAINTING—Fire-department red, striped.

American-LaFrance

STANDARD TYPE 75

Chemical Car

SPECIFICATIONS

Motor—Six cylinders, 5½-inch bore by 6-inch stroke, 105 Horse Power.

Wheel Base—156½ inches.

Wheels—Artillery Type.

Tires—36 x 4, single front, dual rear. Cup cushion.
Pneumatic, or any special type of tires furnished at additional cost.

Lighting System—
Two 12-inch electric headlights.
One 12-inch electric searchlight.

Gasoline Capacity—15 gallons, gravity feed.

Siren Horn—One, hand-operated.

Locomotive Bell—One.

Equipment Box—One, at rear.

Tool Box—One.

Crowbar—One.

Chemical Tanks—Four, each 60-gallon capacity.

Chemical Hose—400 feet of 1-inch chemical hose.

Ladders—
One 20-ft. extension ladder.
One 12-ft. roof ladder with folding hooks.

Pike Pole—One.

Axe—One, fire department standard.

Lanterns—Two, Dietz fire department standard.

Extinguishers—Two, 2½-gallon Babcock.

Bumper—Heavy steel with recoil springs.

ALL NECESSARY OPERATING TOOLS

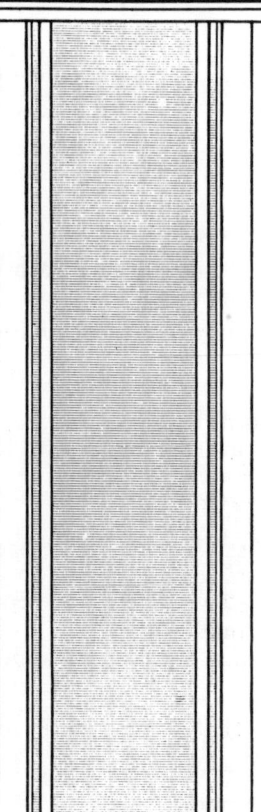

STANDARD TYPE 75

Combination Chemical and Hose Car

SPECIFICATIONS

Motor—Six cylinders, 5½-inch bore by 6-inch stroke, 105 Horse Power.

Wheel Base—156½ inches.

Wheels—Artillery Type.

Tires—36 x 4, single front, dual rear. Cup cushion.
  Pneumatic, or any special type of tires furnished at additional cost.

Lighting System—
  Two 12-inch electric headlights.
  One 12-inch electric searchlight.

Gasoline Capacity—30 gallons, gravity feed.

Siren Horn—One, hand-operated.

Locomotive Bell—One.

Tool Box—One.

Equipment Box—One, at rear.

Crowbar—One.

Hose Capacity—1200 feet of 2½-inch hose.

Chemical Tank—One, 40-gallon capacity.

Chemical Hose—200 feet of ¾-inch chemical hose.

Ladders—
  One 20-ft. extension ladder.
  One 12-ft. roof ladder with folding hooks.

Pike Pole—One.

Play Pipe Cones—Two.

Axe—One, fire department standard.

Lanterns—Two, Dietz fire department standard.

Extinguishers—Two, 2½-gallon Babcock.

Bumper—Heavy steel with recoil springs.

ALL NECESSARY OPERATING TOOLS

American-LaFrance

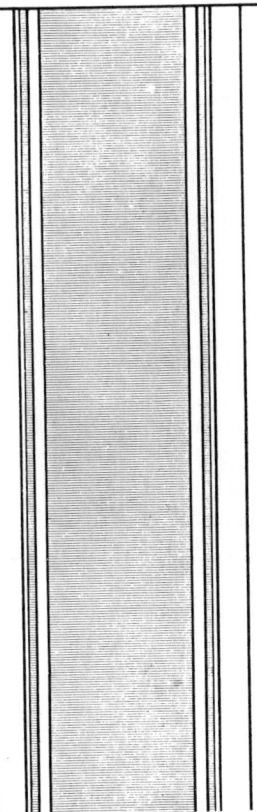

STANDARD TYPE 75

Triple Combination Pumping, Chemical and Hose Car

750 Gallons Capacity — Rotary Gear Pump

SPECIFICATIONS

Motor—Six cylinders, 5½-inch bore by 6-inch stroke, 105 Horse Power.

Wheel Base—156½ inches.

Wheels—Artillery Type.

Tires—36 x 4, single front, dual rear. Cup cushion.
Pneumatic, or any special type of tires furnished at additional cost.

Lighting System—
Two 12-inch electric headlights.
One 12-inch electric searchlight.

Gasoline Capacity—30 gallons, gravity feed.

Siren Horn—One, hand-operated.

Locomotive Bell—One.

Tool Box—One.

Equipment Box—One, at rear.

Crowbar—One.

Hose Capacity—1200 feet of 2½-inch hose.

Chemical Tank—One, 40-gallon capacity.

Chemical Hose—200 feet of ¾-inch chemical hose.

Ladders—
One 20-ft. extension ladder.
One 12-ft. roof ladder with folding hooks.

Suction Hose—Two lengths, each 10 ft. 6 in. long.

Pike Pole—One.

Play Pipe Cones—Two.

Axe—One, fire department standard.

Lanterns—Two, Dietz fire department standard.

Extinguishers—Two, 2½-gallon Babcock.

Bumper—Heavy steel with recoil springs.

American-LaFrance

STANDARD TYPE 40

Combination Chemical and Hose Car

SPECIFICATIONS

Motor—Four cylinders, 5½-inch bore by 6-inch stroke, 75 Horse Power.

Wheel Base—140½ inches.

Wheels—Artillery Type.

Tires—36 x 4, single front, dual rear. Cup cushion.
Pneumatic, or any special type of tires furnished at additional cost.

Lighting System—
Two 12-inch electric headlights.
One 12-inch electric searchlight.

Gasoline Capacity—30 gallons, gravity feed.

Siren Horn—One, hand-operated.

Locomotive Bell—One.

Tool Box—One.

Equipment Box—One, at rear.

Crowbar—One.

Hose Capacity—1200 feet of 2½-inch hose.

Chemical Tank—One, 40-gallon capacity.

Chemical Hose—200 feet of ¾-inch chemical hose.

Ladders—
One 20-ft. extension ladder.
One 12-ft. roof ladder with folding hooks.

Pike Pole—One.

Play Pipe Cones—Two.

Axe—One, fire department standard.

Lanterns—Two, Dietz fire department standard.

Extinguishers—Two, 2½-gallon Babcock.

Bumper—Heavy steel with recoil springs.

ALL NECESSARY OPERATING TOOLS

100

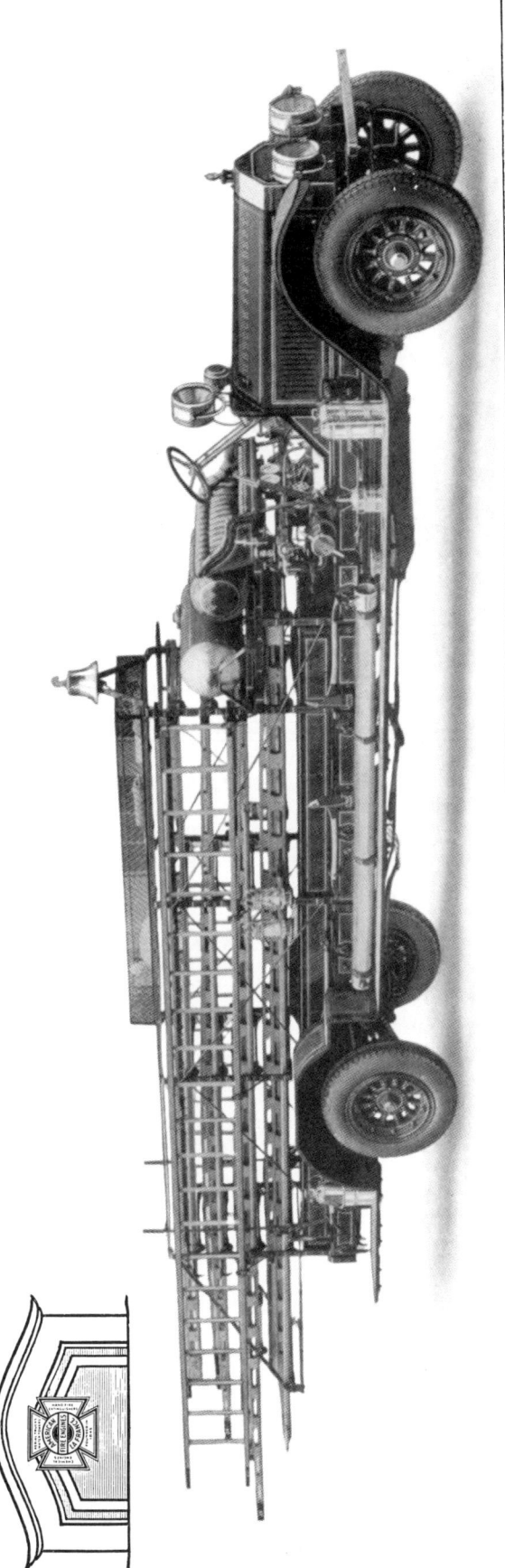

STANDARD TYPE 14
Combination City Service Hook and Ladder Truck
600-Gallon Capacity—Rotary Gear Pump

SPECIFICATIONS

Motor—Six cylinders, 4½-inch bore by 6-inch stroke, 75 Horse Power.
Wheel Base—246½ inches.
Wheels—Artillery Type.
Tires—Single front, dual rear. Cup cushion. Pneumatic, or any special type of tires furnished at additional cost.
Lighting System—
Two 12-inch electric headlights.
One 12-inch electric searchlight.
Gasoline Capacity—20 gallons, gravity feed.
Siren Horn—One, hand-operated.
Locomotive Bell—One, 12-inch.
Chemical Tank—One, 40-gallon capacity.
Hose Body—Capacity of 500 or 1000 feet of 2½-inch hose.
Suction Hose—Two lengths, each 10 ft. 6 in. long.
Chemical Hose—200 feet of ¾-inch chemical hose.
Extinguishers—Two, 2½-gallon Babcock.
Lanterns—Four, Dietz fire department standard.
Axes—Four, fire department standard.
Wall Picks—Two.
Crowbars—Two.
Shovels—Two.
Wire Cutter—One, Boston.
Door Opener—One, Detroit.
Tin Roof Cutter—One, La France.
Pitchforks—Two.
Rubber Buckets—Four.
Pike Poles—Six, assorted lengths
Crotch Poles—Two.
Wire Basket—One.
Tool Box—One.
Ladder Lock—One.
Bumper—One.

ALL NECESSARY OPERATING TOOLS

LADDER EQUIPMENT

One 50-ft. rapid-hoist rope and pulley extension ladder to extend fifty feet, with supporting poles.
One 35-ft. rapid-hoist rope and pulley extension to extend thirty-five feet.
One 28-ft. single ladder.
Two 25-ft. single ladders.
One 24-ft. single ladder.
One 20-ft. single ladder.
One 16-ft. roof ladder.
One 12-ft. roof ladder.
Total ladder equipment, 235 feet.

The OBENCHAIN-BOYER CHEMICAL ENGINE

MOTOR-DRIVEN FIRE APPARATUS

One, Two, Three and Four Tank
Straight Chemical Units

Combination Chemical and
Hose Body Units

Triple Combination Apparatus

IN designing and constructing a Chemical Fire Engine, the fact which must be kept in mind is that it is a machine which is to be used under the most trying circumstances, when the operators will be working under a strain of excitement, when any hitch or delay will prove fatal to the purpose which the engine is designed to serve, that whenever it is called upon, it must, without fail, instantly and vigorously respond, and that, from the nature of the service, it will be subject to the severest kind of usage.

The six points to be looked for in passing judgment on a Chemical Fire Engine are as follows:
Absolute sureness of action.
Speed and ease of operation.
Provision for dissolving the soda.
Provision for sealing the acid.
Security against overcharging.
Grade of material used, standard of workmanship, general appearance and finish.

With the above points in mind we invite you to the consideration of the Obenchain-Boyer Chemical Fire Engine. Reference to the accompanying drawing of the interior of the tank will render these points discussed clear.

OPERATION OF ENGINE

When this type of engine is drawn to the scene of a fire there is but one thing necessary to set it in operation, that is, to pull, through a quarter turn, the operating lever on top of the tank. This lever actuates the bottle-breaking mechanism and within ten seconds from the time it is operated the pressure gauge will register from 140 to 180 pounds pressure.

BOTTLE BREAKING DEVICE

The device for breaking the bottle is so simple that it will, in all cases, positively effect its purpose. The bottle holder is so constructed that it retains all the broken glass and not a particle of it can fall into the tank.

IN presenting the Obenchain-Boyer line of motor driven equipment, special attention is called to the fact that, contrary to ordinary usage, when a body is provided for any chassis the entire equipment is practically built from the ground up, and not simply bolted on to the frame. This insures a compact fire-fighting unit and reduces to a minimum the unusually heavy wear and tear to which this class of apparatus is subjected.

The Pacific Fire Extinguisher Company is prepared to furnish any type of body from a straight chemical unit to a triple combination for any make of automobile from three quarters of a ton to two tons, quotations being based upon a complete equipment excluding nothing but the chassis itself. The following specifications will, therefore, in a general way cover practically all the types of apparatus most commonly in use.

SPECIFICATIONS

CHEMICAL TANKS—Standard, patented, OBENCHAIN-BOYER 35 gallon capacity tanks. Tanks made of best hearth flange steel, highly finished. Heads swaged in and side seams riveted and copper brazed. Tanks have been chemically treated to prevent corrosion. Each tank has a clean-out valve, easily accessible. All tanks securely fastened to platform with heavy brackets and steel rods.

AGITATOR—Flexible, to dissolve soda. Agitator in each tank, working on bottom of tank. Made of solid steel rod covered with coppered coil spring.

SAFETY FLANGE—Safety flange in each tank, absolutely preventing overcharging and excessive pressure.

GAUGES—One (1) on each tank. Calibrated to 400 pounds.

BY-PASS—One (1) hose connection for attaching fire hose to fill tanks or to run water through chemical hose, using the latter as a deadener. Connection fitted with cap.

HOSE BASKET—One (1) hose basket made of high grade flat woven wire. Rectangular in shape, brass edged, so as to allow hose to play out easily. Very substantially built. Basket is supported by heavy brackets. Outlet of chemical tank connects to the chemical hose on the interior of basket.

CHEMICAL HOSE—150 feet of ¾ inch Underwriters cotton web, rubber lined hose coupled in 50 ft. sections.

NOZZLE—One (1) Eccentric brass Underwriters shut-off nozzle. This nozzle has a five sixteenth (5-16) inch tip. Nozzle equipped with a lever handle.

CHEMICAL BOX—Metal chemical box, suitably located on running board. Capacity of two complete charges.

CHARGES—Three (3) complete charges furnished with each tank.

BODY FRAME—Body frame made of heavy straight and close grain, hard wood, five years seasoned. Solid frame plates from dash to extreme rear, well cross braced, strongly reinforced with hand forged wrought iron braces. Frame is securely fastened to chassis, steel frame, with extra large flat head steel bolts. There are four (4) body cross bars; end bars extend outside of body 6 inches, providing for corner braces and are 2 inches x 6 inches well seasoned, straight and close grained hard wood. The other two are 2 inches x 4 inches same kind of wood, spaced at equal distances between the front and rear bars.

HOSE BODY—Capacity of 800 to 1500 feet, as may be specified, of Double Jacket Fire Hose. Hose body is of steel exterior and a wood slat interior. Body posts are gained in cross bars, which carry the posts and are bolted to the bars by two (2) 5-16 inch bolts. Bottom of body covered with 4½ inch x ⅝ inch slats of well seasoned straight grained hard wood, having ½ inch space between slats to insure circulation of air. Rear end of slats covered with angle iron. Each slat attached to each cross bar with two (2) large flat head wood screws. Sides are covered with well seasoned hard wood 2 inch slats, ½ inch space between slats allows the circulation of air on the sides. This prevents deterioration of hose. These slats are attached to body posts with flat head screws. Front end of body has a hard wood front supported and connected to the sides with 3 inch angle iron, extending full depth of bed. Exterior of front covered with sheet metal the same as the sides, lapped on the front corners. Sides covered with heavy sheet steel, attached to body posts with rivets. Front end edge of body is covered with ¾ inch x ½ inch round iron. Running from the front end to the extreme rear, on the edges of both sides, is 2 inch channel steel, completely covering the edges of the steel exterior and the slat interior, as well as the body posts. This channel steel is securely fastened to body posts with long coach screws. At each end of the body are ladder brackets and body braces combined, made of 1½ inch x ½ inch steel, hand forged, securely bracing the body.

BRASS RAILINGS—Brass railings, high grade tubular, outside diameter one (1) inch extended from front of body to rear step. Substantially supported with brass stanchions. All brass railings are highly polished. A brass cross bar, of 1-inch brass rail connects rear ladder supports, conveniently located for hand hold. This insures proper bracing of body sides, when laying out hose on turns.

RUNNING BOARDS—Running boards made of well seasoned hard wood. Extend between the front and rear fenders. Well braced and securely bolted to body frame; covered with corrugated rubber matting, interior edges bound by ¾ inch brass and on outer edge by angle iron, which protects edge of board and binds matting.

FENDERS—Front fenders are standard equipment of chassis, securely fastened to running board. Rear fenders are drum pattern made of heavy sheet metal, securely bolted to running boards and rear step.

REAR STEP—Rear step directly connected to body frame, securely braced with wrought iron braces. Extends entire width from out to out of fenders. 16 inches wide, 66 inches long. Made of thoroughly seasoned hard wood. Side extensions directly fastened to fenders. Covered with corrugated rubber matting, interior edge protected with brass nozing and exterior edges protected with angle iron same width as thickness of step.

BODY SKIRTS—Body skirts extend entire length of apparatus between the front and rear fenders, entire width of rear, completely covering space between steps and body and fenders and body. Substantially built of best heavy steel, well supported, and all connections securely fastened.

LADDER SUPPORTS—Ladder supports made of 1½ inch x ½ inch hand forged steel, securely braced. Interior of ladder locks covered with high grade leather. Ladders securely locked with keys, which are conveniently fastened to ladder supports with chains. Ladder supports so located that the weight of ladders is proportionately carried on the truck to prevent vibration.

LADDERS—List of ladders as follows:
  One 24-foot extension, rapid hoist ladder with solid sides and safety locks.
  One 12-foot roof ladder, with folding hooks.
  One 10-foot pike pole.
  Sides of ladders made of 5 year seasoned, straight grain pine, with hickory rungs, filled and varnished, ends finished in black, also having heel plates.

SEAT—Standard automobile seat. Back and sides are all one piece wood. Exterior completely covered with steel, and inside well braced with steel angles. Corners and sides rounded to make seat easily accessible. Spring cushion seat well upholstered. Back padded and completely covered with high grade upholstering.

HAND HOLDS—Hand holds on each side of seat. Also one hand hold on each side of dash.

FLOOR BOARDS—Floor boards covered with corrugated rubber matting, edges protected with brass. Each board being separately covered so that boards can be removed without removal of matting.

EXTINGUISHERS—Two (2) 2½ gallon heavy polished copper hand fire department extinguishers.

EXTINGUISHER HOLDERS—Two extinguisher holders, or cups, conveniently located on running boards, having up-rights containing leather strap and buckle, which securely fastens the extinguisher into its holder.

LANTERNS—Two Dietz standard firemen's lanterns, conveniently fastened in holders. One on each side of apparatus.

AXE—One Fireman's pick head axe, conveniently located, with handle holder and axe scabbard.

CROWBAR—One heavy 36-inch crowbar. Located in holders on the rear step.

ALARM SYSTEM—One Sterling hand or electric siren, located on right hand side of cowl dash.

PAINTING—Entire outfit painted in Sagamore Red with gold and black stripes. Auto finished with acid proof varnish.

LETTERING—According to the request of purchaser.

TWO TANK STRAIGHT CHEMICAL UNIT

The usual combination of chemical tank and hose body, as follows:
- ¾ ton truck, 1 35-gal. Chemical Tank, 800' hose body.
- ¾ ton truck, 2 35-gal. Chemical Tanks, 600' hose body.
- 1 ton truck, 1 35-gal. Chemical Tank, 1000' hose body.
- 1 ton truck, 2 35-gal. Chemical Tanks, 800' hose body.
- 1½ ton truck, 1 35-gal. Chemical Tank, 1200' hose body.
- 1½ ton truck, 2 35-gal. Chemical Tanks, 1000' hose body.
- 2 ton truck, 1 35-gal. Chemical Tank, 1800' hose body.
- 2 ton truck, 2 35-gal. Chemical Tanks, 1500' hose body.

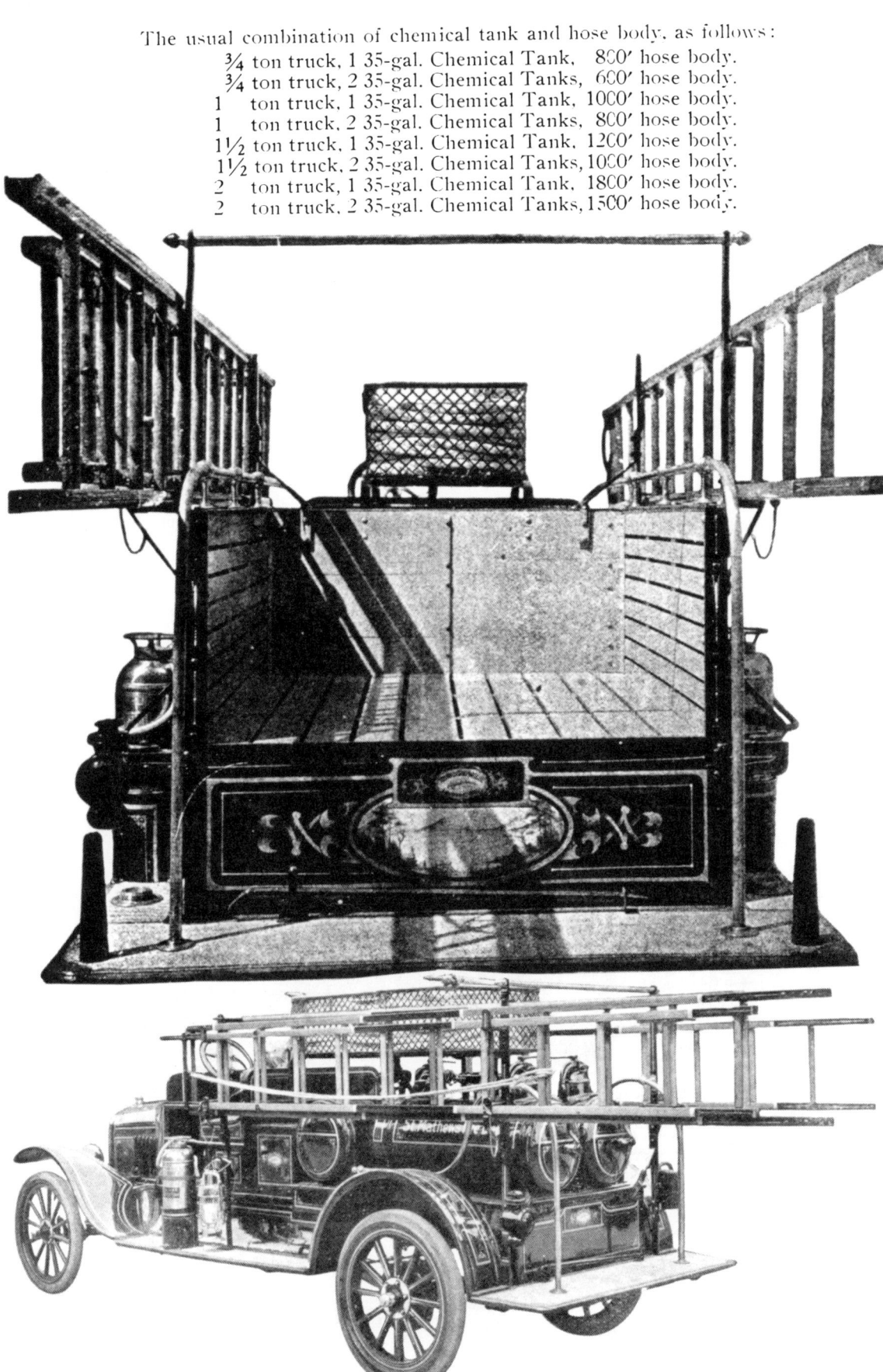

THREE TANK STRAIGHT CHEMICAL UNIT

TYPE OF STRAIGHT CHEMICAL UNIT

FOUR TANK TYPE OF STRAIGHT CHEMICAL UNIT

COMBINATION DOUBLE TANK AND HOSE BODY UNIT

COMBINATION SINGLE TANK AND HOSE BODY UNIT

COMBINATION SINGLE TANK AND HOSE BODY UNIT

COMBINATION DOUBLE TANK AND HOSE BODY UNIT

COMBINATION TRIPLE TANK AND HOSE BODY UNIT

COMBINATION FOUR TANK AND HOSE BODY UNIT

COMBINATION TRUCK AND CHEMICAL

Triple Combination Apparatus

THE Hale Rotary motor driven pump, which is used exclusively in connection with all Obenchain-Boyer triple combination units, is available in four capacities.

Type "A" of Ford units, capacity 200 gallons.
Type "B" 225-250 gallons capacity.
Type "C" 350 gallons capacity.
Type "D" 500 gallons capacity.

In discussing the relative merits and available pressures of comparatively small discharge pumps of this type, a considerable amount of difficulty is experienced among most fire officers concerning the pressure question, this being a very difficult matter in connection with a commercial chassis. Therefore, instead of attempting to tabulate them on the basis of pressure, they are tabulated on the stream basis, which figures out as follows:

Through 1 500 ft. line of 2½" Standard Hose with 1" Nozzle, 80 feet.
Through 1 700 ft. line of 2½" Standard Hose with ¾" Nozzle, 70 feet.
Through 2 250 ft. lines of 2½" Standard Hose with ¾" Nozzle, 65 feet.
Through 3 200 ft. lines of 2½" Standard Hose with ⅝" Nozzle, 50 feet.
Through 1 1000 ft. line of 2½" Standard Hose with ¾" Nozzle, 60 feet.

Care should be exercised in figuring pump capacity and motor horse power, the Underwriters requiring 1 horse power for each 10 gallons of water per minute at 120 pounds pressure based on the S. A. E. rating of the motor.

STANDARD TYPE OF TRIPLE COMBINATION

STANDARD TYPE OF TRIPLE COMBINATION

BOYER
Built-to-Order
Motor Fire Apparatus
includes:

Combination Hose and Chemical Trucks
Motor Driven Hook and Ladder Trucks
Motor Driven Salvage Trucks

Motor Driven Pumpers
Hand Pulled Chemical Engines
Hand Pulled Hook and Ladder Trucks

When you buy Boyer Specially Built Apparatus you select the chassis and every part of the equipment, and the complete apparatus is constructed to fill your own particular needs—built to your order.

Double Tank Chemical

Triple Tank Chemical

This method, the Obenchain-Boyer way, makes it possible for you to procure the modern and thoroughly dependable equipment your city should have without the heavy expense often associated with the purchase of apparatus.

Pick out a standard commercial chassis of proven dependability, and we equip it specially as a combination chemical, pumper, hook and ladder, city service car, or in any way your requirements demand. In other words, from headlights to step, it is equipped as you want and need it.

Combination Hose and Chemical

Triple Combination, Hose, Chemical and Pump

It will pay you to know all about this method, and, whether or not you are planning purchases of equipment at this time, we will gladly mail full particulars upon request. Just drop us a postal card.

Address:

The Obenchain-Boyer Co., Logansport, Ind.

"Pirsch" Two Tank Chemical Car—Ford One-Ton Chassis

SPECIFICATIONS

Rear Steps—Of hardwood, hung from rear end of chassis by hand forged braces, covered with linoleum, rubber matting or aluminum.

Railings—Heavy iron pipe railings running from front down to rear footboard and fastened there by means of iron sockets.

Basket—One wire basket behind seat for boots, coats, etc.

Chemical Tanks—2 35-gallon "PIRSCH-CHAMPION" seamless pressed steel tanks.

(Also built in three and four tank units with double chemical lines.)

Chemical Tank Equipment—Four acid bottles; two acid bottle holders for running board; two soda bags; one wrench; two gauges; one 2½" hose connection; one set double piping.

Ladders—1 12' solid side roof ladder with folding hooks attached; 1 20' solid side extension ladder, rope hoist. Ladders are carried in strong iron holders. One ladder on each side. Ladders finished in natural color of the wood, ends black.

Hose Reel—One automatic hose reel. (Can furnish hose bracket if desired.)

Chemical Hose—150' ¾" chemical hose, fitted with heavy brass couplings, coupled in 50' lengths.

Chemical Nozzle—One chemical shut-off nozzle.

Extinguishers—Two 3-gallon fire extinguishers in iron holders, carried on running board.

Marine Lights—Two marine lights on rear step.

Lanterns—Two Dietz Fire Department lanterns in holders. (Or one hand electric light if preferred.)

Bell—One 10" locomotive bell, mounted on basket, dash or in front of radiator. (Hand or electric siren optional if desired in place of bell.)

Axe—One pick back fire axe in holders.

Crow Bar—One crow bar in holders.

Pike Pole—One 8' pike pole in holders.

Nozzle Plugs—Two hardwood nozzle plugs on rear steps.

Rear Fenders—Of suitable size, attached to side and rear steps.

Side Steps and Driver's Footboards—Covered with linoleum, rubber matting or aluminum, with binding strips.

Painting and Lettering—Painting is done in first class manner with best wearing coach colors and varnishes. Painted in any color desired, striping, lettering and decorations done in gold leaf with appropriate shadings. Lettering to order.

This equipment also built with three and four chemical tanks.

EVERYTHING FOR THE FIRE DEPT.

"Pirsch" Combination Chemical and Hose Car—Ford One Ton Chassis

SPECIFICATIONS
Combination Chemical Hose Car Equipment.

Hose Body: Constructed of steel panels, angle steel frames, top, bottom and rear, with hardwood sills. Floor of body 4" hardwood slats with ½" space between each to allow for the ventilation of the hose. Capacity 1000' 2½" regulation fire hose.

Rear Steps: Of hardwood, hung from rear end body by hand forged braces, covered with linoleum and bound on edges with 1" half oval iron. Under body and chassis frame there is a hardwood tool box.

Railings: Heavy iron pipe railings running from front end of body down to rear footboard and fastened there by means of iron sockets.

Driver's Seat: Of suitable size for two men, upholstered, with cushion, stuffed with curled hair. Seat equipped with hand grabs on either side.

Chemical Tank: 1 35-gallon "Pirsch Champion" seamless pressed steel tank.

Chemical Tank Equipment: Two acid bottles; one acid bottle holder for running board; one soda bag; one wrench; one gauge; one 2½ hose connection.

Ladders: 1-12' solid side roof ladder with folding hooks attached; 1-20' solid side extension ladder, rope hoist. Ladders are carried on hose box in strong iron holders. One ladder on each side of body. Ladders finished in natural color of the wood, ends black.

Hose Basket: One perforated steel hose basket, with wooden slatted bottom and rollers on top, mounted over body. (Can furnish automatic hose reel if desired.)

Chemical Hose: 150' ¾" chemical hose, fitted with heavy brass couplings, coupled in 50' lengths.

Chemical Nozzle: One chemical shut-off nozzle.

Extinguishers: Two 3-gallon fire extinguishers in iron holders, carried on running-board.

Lanterns: Two Dietz Fire Department lanterns in holders.

Bell: One 8" locomotive bell, mounted on basket, dash or in front of radiator.

Axe: One pick back fire axe in holders.

Crow Bar: One crow bar in holders.

Pike Pole: One 8' pike pole in holders.

Nozzle Plugs: Two hardwood nozzle plugs on rear step.

Rear Fenders: Of sufficient size, attached to side and rear steps.

Side Steps and Driver's Footboards: Covered with linoleum with binding strip.

Painting and Lettering: Painting is done in first class manner with the best wearing coach colors and varnishes. Painted in any color desired, striping, lettering and decorations done in gold leaf with appropriate shadings. Lettering to order.

This equipment also built with double chemical tanks.

PETER PIRSCH & SONS CO.

"Pirsch" Combination Chemical Hose and Pumping Car—Ford One Ton Chassis

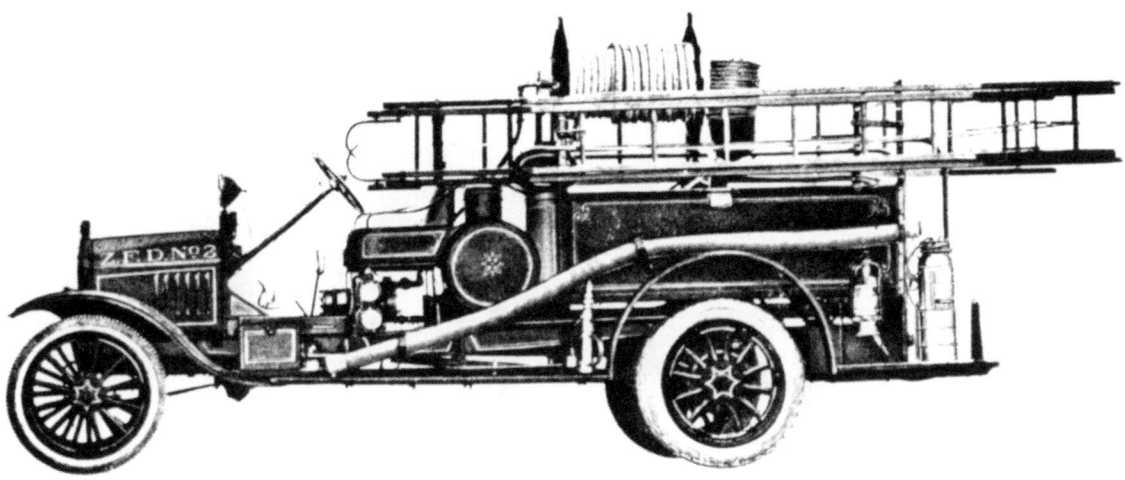

SPECIFICATIONS

Triple Combination Chemical Hose and Pumper Equipment--Either
Soda and Acid or Booster Tank

Hose Body—Constructed of steel panels, angle steel frames, top, bottom and rear, with hardwood sills. Floor of body of 4" hardwood slats with ½" space between each to allow for the ventilation of the hose. Capacity 1000 feet of 2½" regulation fire hose.

Rear Step—Of hard wood, hung from end of body by hand forged braces covered with either rubber, linoleum or aluminum matting and metal edging. Under body and chassis frame there is a hardwood tool box with hinges and lock.

Railings—Heavy brass 1¼" outside diameter running from front end of body down to rear footboard and fastened there by means of iron sockets.

Driver's Seat—Of suitable size for two men, upholstered, with cushion stuffed with curled hair. Seat equipped with hand grabs on each side.

Chemical Tank—Option of either one 35 or 40-gallon "Pirsch" seamless steel chemical tank complete with two acid bottles, one acid bottle holder for side step, one set piping with three valves, by-pass from fire pump into tank, soda bag, one soda canister, wrench, tipping wheel and pressure gauge. All fittings on tank of best grade red brass. Tank and all fittings tested to 350 pounds hydrostatic pressure; or
One 50-gallon galvanized iron inexhaustible tank, with large opening in top and twelve galvanized iron buckets for refilling tank when exhausted. This tank is arranged so that bi-carbonate soda may be mixed with water and pumped through chemical hose, as no pressure is used in the tank itself, it may be refilled while in operation.

Pump—The pump is of the rotary type with phosphor bronze rotars, alloy steel shafts, annular ball bearings, patented bronze stuffing boxes.

Drive Unit—The drive unit consists of an oil-tight case carrying nickel steel hardened gears on annular ball bearings. The sliding gear is operated by a lever, which is on the outside of case convenient to the driver. This drive unit case is bolted to the pump head in such a manner that it is practically part of the pump.

Pump Equipment—2-10' lengths suction hose with brass couplings; 1 brass strainer; 2 outlet valves with brass caps; 2 brass caps for suction openings; 1 pressure gauge; 1 chemical connection, for connecting pump with chemical tank; 1 reducer for pumping off hydrant.

Cooling—Fresh water line with shut off valve from pump to motor.

Ladders—1-12' solid side roof ladder with folding hooks attached; 1-20' solid side extension ladder, rope hoist. Ladders are carried on hose box in strong iron holders. One ladder on each side of body. Ladders finished in natural color of the wood, ends black. Made of Washington fir rails and second growth ash rungs.

Hose Reel—One automatic hose reel. Capacity 250 feet ¾" chemical hose, or if preferred a steel hose basket will be furnished.

Chemical Hose—150' ¾" chemical hose, fitted with heavy brass couplings, coupled in 50' lengths.

Chemical Nozzle—One chemical shut-off nozzle.

Extinguishers—Two 3-gallon fire extinguishers in iron holders carried on running-board. Either sealed or break bottle type with straps and shut-off nozzle.

Lanterns—Two Dietz Fire Department lanterns in holders.

Bell—One 10" locomotive bell, mounted on basket, dash or in front of radiator.

Axe—One pick back fire axe in holders.

Crow Bar—One crow bar in holders.

Pike Pole—1-8' pike pole in holders.

Nozzle Plugs—Two hardened nozzle plugs on rear step.

Rear Fenders—Of sufficient size, attached to side and rear steps.

Rubber Mat—One for driver's floor boards.

Side Steps and Driver's Footboards—Covered with linoleum with binding strip.

Painting and Lettering—Painting is done in first class manner with the best wearing coach colors and varnishes. Painted in any color desired, striping, lettering and decorations done in gold leaf with appropriate shadings. Lettering to order.

EVERYTHING FOR THE FIRE DEPT.

"Pirsch" Chemical Car Equipment
For Any Chassis
(BUILT IN TWO, THREE AND FOUR TANK UNITS)

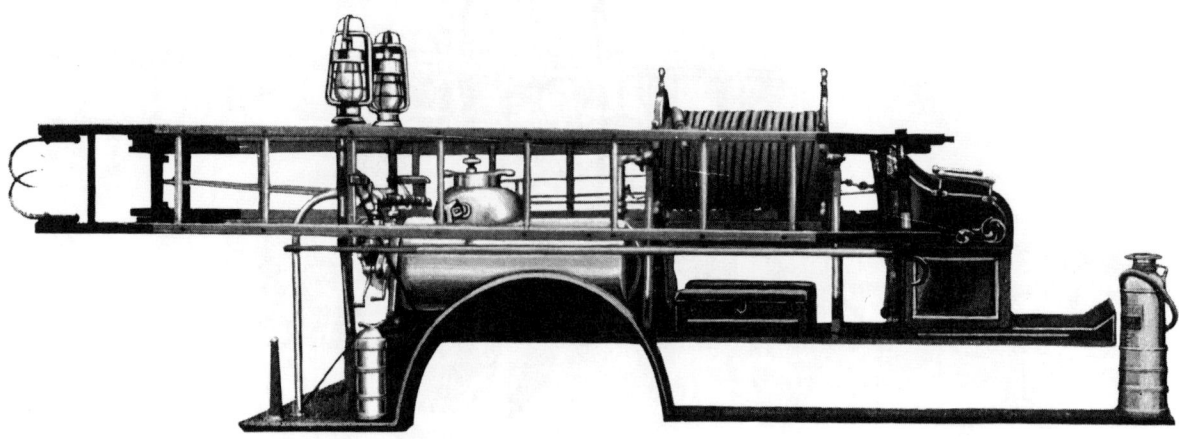

SPECIFICATIONS

Rear Steps—Of hardwood, hung from rear end of chassis by hand forged braces, covered with linoleum, corrugated rubber matting or aluminum, and bound on edges with 1" half oval iron. Under chassis frame there is a hardwood tool box.

Railings—Heavy railings running from front down to rear footboard and fastened there by means of sockets.

Driver's Seat—Of suitable size for two men, upholstered, with cushion, stuffed with curled hair. Seat equipped with hand grabs on either side.

Chemical Tank—2 40-gallon (or greater capacity) "Pirsch Champion" or "Pirsch Standard" chemical tanks.

(Also built with three or four tanks and double chemical lines.)

Chemical Tank Equipment—Four acid bottles; two acid bottle holders for running board; two soda bags; one wrench; two gauges; one 2½" hose connection; one set double piping and overflow valves.

Chemical Charges—Two complete charges for each tank.

Ladders—One 12' solid side roof ladder with folding hooks attached; one 20' solid side extension ladder, rope hoist. Ladders are carried in strong iron holders. One ladder on each side. Ladders finished in natural color of the wood, ends black.

Hose Reel—One automatic hose reel.

Chemical Hose—150' ¾" chemical hose, fitted with heavy brass couplings, coupled in 50' lengths.

Chemical Nozzle—One chemical shut-off nozzle.

Extinguishers—Two 3-gallon Fire Dept. extinguishers in iron holders, carried on running board.

Lanterns—Two Dietz Fire Department lanterns in holders.

Bell—One 10" locomotive bell, mounted on dash or in front of radiator.

Axe—One pick back fire axe in holders.

Crow Bar—One crow bar in holders.

Pike Pole—One 8' pike pole in holders.

Rear Fenders—Of sufficient size attached to side and rear steps.

Side Steps and Driver's Footboards—Covered with linoleum, corrugated rubber matting, or aluminum, with binding strips.

Painting and Lettering—Painting is done in first-class manner with the best wearing coach colors and varnishes. Painted in any color desired, striping, lettering and decorations done in gold leaf with appropriate shadings. Lettering to order.

Note—If desired, three or four chemical tanks may be specified in place of two.

"PIRSCH" SINGLE TANK, CLASS "2-C"

Combination Chemical and Hose Body Equipment

(Especially Designed for Ford One Ton Chassis)
May Be Used on Other Chassis

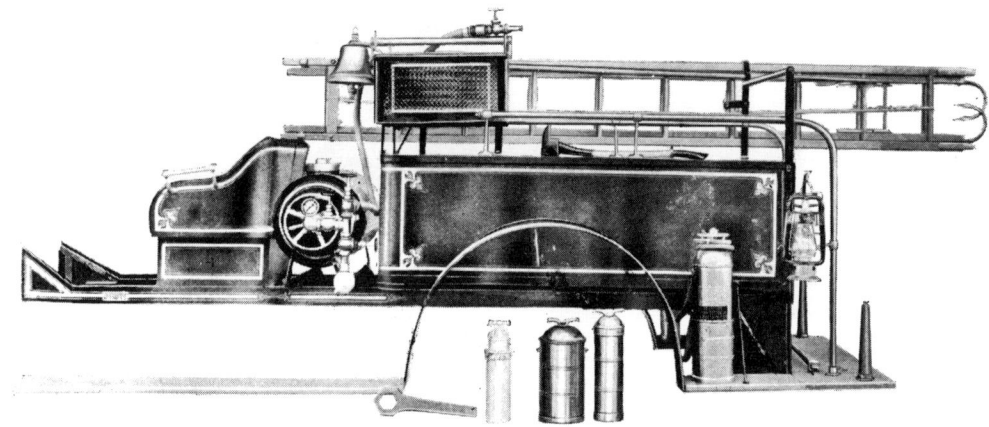

SPECIFICATIONS

Hose Body—Constructed of extra special heavy pickeled steel panels, angle steel frames, top, bottom and rear, with hardwood sills. Floor of body of 4" hardwood slats with ½" space between each to allow for ventilation of the hose; floor of body removable over transmission, differential, etc. Front corners of body nicely curved, interior of body strongly braced by means of steel braces, securely bolted to cross sills on bottom of body and riveted to panels. Capacity 1000' 2½" hose.

Rear Step—Of hardwood, hung from rear end of body by hand forged braces. Step is 16" or deeper if preferred and as wide as the outside width of rear mudguards, covered with linoleum, corrugated rubber matting or aluminum, and bound on edges with metal binding. Directly in front of rear step and under body and chassis frame there is a hardwood tool box with spring drop door and lock.

Railings—Heavy pipe railings running from front end of body down to rear footboard and fastened there by means of sockets. Fastened to body by means of stanchions, front ends of railings fitted with ornamental acorns. There is also a rail which serves as a hand hold for men riding on the rear step.

Driver's Seat—Of suitable size for two men, upholstered, with spring back and cushions, stuffed with hair. Seat equipped with hand grabs on either side.

Chemical Tank—One 35 or 40-gallon steel "Pirsch Champion" tank. This tank is operated by revolving on trunions by means of a tipping wheel. The piping consists of three valves, a 2½" hose connection, chemical hose connection and a 2½" pressure gauge. The threads on the 2½" hose connection are cut to fit the city thread so that the tank may be refilled through this connection, or fresh water thrown on the fire through chemical hose. Tank is mounted directly behind seat. All piping properly threaded, sweated and soldered.

Chemical Tank Equipment—Two acid bottles, one acid bottle holder for running board, one soda bag and one wrench.

Chemical Charges—Two complete charges.

Ladders—One 12' solid side roof ladder with folding hooks attached; one 24' solid side extension ladder, rope hoist. Ladders are carried on hose box in strong iron holders. One ladder on each side of body, or both on one side, just as desired. Ladders furnished in natural color of wood, ends black.

Hose Basket—One steel hose basket with wooden slatted bottom, mounted over body.

Chemical Hose—150' ¾" chemical hose, fitted with heavy brass couplings, coupled in 50' lengths.

Chemical Nozzle—One chemical shut-off nozzle.

Extinguishers—Two 3-gallon Fire Dept. extinguishers in iron holders. Carried on running boards.

Lanterns—Two Dietz Fire Dept. lanterns in holders.

Bell—One 8" locomotive bell mounted on basket, dash or front of radiator.

Axe—One pick back fire axe in holders.

Crow Bar—One crow bar in holders.

Nozzle Plugs—Two hardwood nozzle plugs on rear steps.

Fenders—For rear wheels of sufficient size (unless furnished).

Side Steps and Driver's Footboards—To be covered with linoleum, corrugated rubber matting or aluminum, with metal binding strip.

Painting and Lettering—Painting to be done in first-class manner, with best wearing coach colors and varnishes. Painted in any color desired; striping, lettering and decorations done in gold leaf with appropriate shadings. Lettering to order.

PETER PIRSCH & SONS CO.

"PIRSCH" SINGLE TANK
Combination Chemical and Hose Body and Equipment

For Any Commercial Chassis

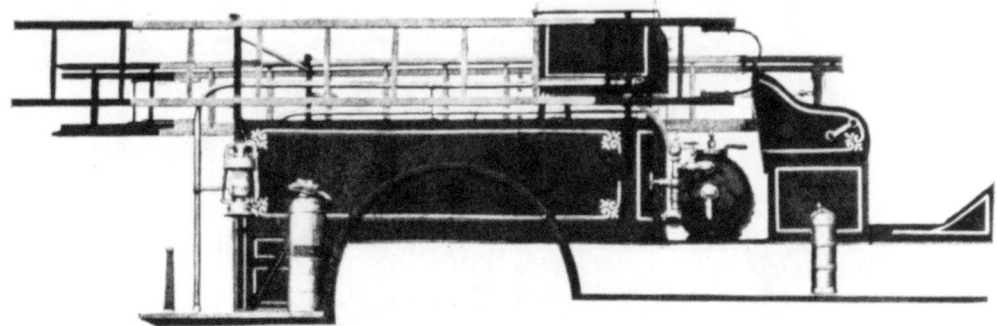

SPECIFICATIONS

Hose Body—Constructed of extra special heavy pickeled steel panels, angle steel frames, top, bottom and rear, with hardwood sills, floor of body of 4" hardwood slats with ½" space between each to allow for the ventilation of the hose, floor of body removable over transmission, differential, etc. Front corners of body nicely curved, interior of body strongly braced by means of steel braces, securely bolted to cross sills on bottom of body and riveted to panels. Capacity as desired.

Rear Step—Of hardwood, hung from rear end of body by hand forged braces. Step is 16" or deeper if preferred and as wide as the outside width of rear mudguards, covered with either linoleum, corrugated rubber matting or aluminum and bound on edges with metal binding. Directly in front of rear step and under body and chassis frame there is a hardwood tool box, with spring drop door and lock.

Railings—Heavy railings running from front end of body down to rear footboard and fastened there by means of sockets. Fastened to body by means of stanchions, front end of railings fitted with ornamental acorns. There is also a rail which serves as a hand hold for men riding on the rear step.

Driver's Seat—Of suitable size for two men, upholstered in genuine leather with spring back and cushions, stuffed with curled hair. Seat equipped with hand grabs on either side.

Chemical Tank—One 35 or 40-gallon "Pirsch Champion" or "Pirsch Standard" tank. The piping consists of three valves, a 2½" hose connection, chemical hose connection and a 2½" pressure gauge. The threads on the 2½" hose connection are cut to fit the city thread, so that the tank may be refilled through this connection or fresh water thrown on the fire through chemical hose. All piping, tipping wheel, caps, etc., are of the best grade red brass, polished. All joints sweated, threaded and soldered. Tank is mounted directly behind seat.

Chemical Tank Equipment—Two acid bottles, one acid bottle for running board, one soda bag and one wrench.

Chemical Charges—Two complete charges.

Ladders—One 12' Pirsch Patent trussed roof ladder with folding hooks attached; one 24' Pirsch Patent trussed extension ladder, rope hoist. Ladders are carried on hose box in strong iron holders. One ladder on each side of body, or both on one side just as desired. Ladders furnished in natural color of wood, ends black.

Hose Basket—One perforated steel hose basket with wooden slatted bottom and rollers on top, mounted over body. (Can furnish automatic hose reel if desired).

Chemical Hose—150' ¾" chemical hose, fitted with heavy brass couplings, coupled in 50' lengths.

Chemical Nozzle—One chemical shut-off nozzle.

Extinguishers—Two 3-gallon Fire Dept. extinguishers in holders. Carried on running boards.

Lanterns—Two Dietz Fire Dept. lanterns in holders.

Bell—One 10" locomotive bell mounted on basket, dash or front of radiator. —or—

Siren Horn—One hand or electric siren horn.

Searchlight—One 11" dash electric swivel searchlight.

Axe—One pick back fire axe in holders.

Crow Bar—One crow bar in holders.

Pike Pole—One 8' pike pole in holders.

Nozzle Plugs—Two hardwood nozzle plugs on rear step.

Torches—Option of either two brass ornamental torches on rear standards or one Pirsch patent hose shut-off and door opener on side step.

Fenders—For rear wheels of sufficient size (unless furnished).

Side Steps and Driver's Footboards—To be covered with corrugated rubber matting, linoleum or aluminum with metal binding strip.

Painting and Lettering—Painting to be done in first-class manner, with best wearing coach colors and varnishes. Painted in any color desired; striping, lettering and decorations done in gold leaf with appropriate shadings. Lettering to order.

EVERYTHING FOR THE FIRE DEPT.

"PIRSCH" DOUBLE TANK
Combination Chemical and Hose Body and Equipment

For Any Commercial Chassis

SPECIFICATIONS

Hose Body—Constructed of extra special heavy gauge pickeled steel panels, angle steel frames, top, bottom and rear, with hardwood sills. Floor of body of 4" hardwood slats with ½" space between each to allow for the ventilation of the hose, floor of body removable over transmission, differential, etc. Front corners of body nicely curved, interior of body strongly braced by means of steel braces, securely bolted to cross sills on bottom of body and riveted to panels. Capacity as desired.

Rear Step—Hardwood, hung from rear end of body by hand forged braces. Step is 16" or deeper if preferred and as wide as the outside width of rear mudguards, covered with either linoleum, corrugated rubber matting or aluminum and bound on edges with metal binding. Directly in front of rear step and under body and chassis frame there is a hardwood tool box, with spring drop door and lock.

Railings—Heavy railings running from front end of body down to rear footboard and fastened there by means of sockets. Fastened to body by means of stanchions, front end of railings fitted with ornamental acorns. There is also a rail in the rear of body, between torch or ladder standards, which serves as a hand hold for men riding on the rear step.

Driver's Seat—Of suitable size for two men, upholstered in genuine leather with spring back and cushions, stuffed with curled hair. Seat equipped with hand grabs on either side.

Chemical Tanks—Two 35 or 40-gal. "Pirsch-Champion" or "Pirsch Standard" type chemical tanks. Interior of tanks tinned to prevent corrosion; all joints properly sweated, threaded and soldered. All fittings are of best grade red brass. Piping consists of six valves, two pressure gauges, 2½" hose connection and fittings, and overflow valves.

Chemical Tank Equipment—Four lead acid receptacles, two brass acid receptacle holders for side step, two soda bags, wrench, etc.

Chemical Charges—Two complete charges for each tank.

Ladders—One 24' Pirsch patent trussed extension ladder, rope, hoist. One 14' patent trussed roof ladder with folding hooks, carried on sides of body in forged iron holders.

Hose Reel—One automatic hose reel carried on top of body. Capacity 300' (or basket if desired.)

Chemical Hose—150' ¾" 4-ply chemical hose with brass couplings, coupled in 50' lengths.

Chemical Nozzle—One chemical shut-off nozzle, ¾" size.

Extinguishers—Two 3-gallon Fire Dept. Extinguishers, carried in holders on side steps.

Lanterns—Two Dietz Fire Dept. lanterns carried in holders.

Bell—One 10" locomotive type bell.

Siren Horn—Either one electric or one hand siren horn. —or—

Searchlight—One 11" dash electric swivel searchlight.

Axes—Two pick back fire axes in holders.

Crow Bar—One steel crow bar in holder.

Pike Pole—One 8' pike pole in holders.

Torches—Two brass torches on rear ladder standards, or one Pirsch patent hose shut-off and door opener on side step.

Fenders—For rear wheels of sufficient size (unless furnished).

Side Steps and Driver's Footboards—To be covered with corrugated rubber matting, linoleum or aluminum, with metal binding strip.

Painting and Lettering—Painting to be done in first-class manner, with best wearing coach colors and varnishes. Painted in any color desired; striping, lettering and decorations done in gold leaf with appropriate shadings. Lettering to order.

PETER PIRSCH & SONS CO.

"Pirsch" Triple Combination Chemical, Hose and Pump Body with Equipment

For Any Commercial Chassis

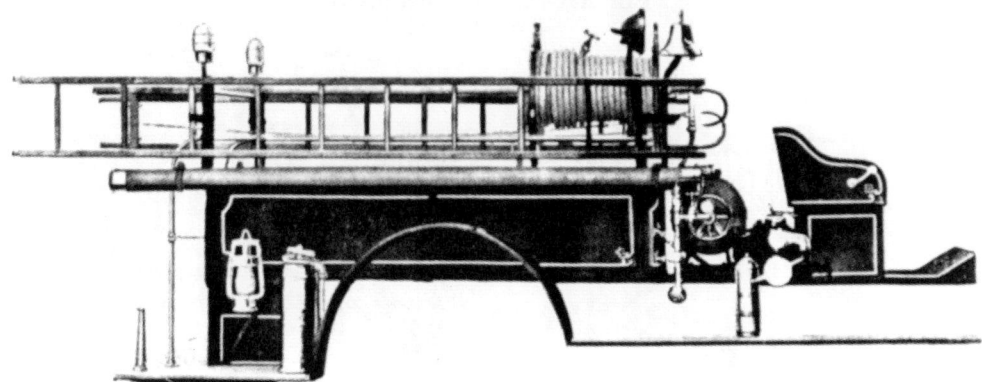

SPECIFICATIONS

Hose Body—Constructed of extra special heavy pickeled steel panels, angle steel frames, top, bottom and rear, with hardwood sills. Floor of body 4" hardwood slats with ½" space between each to allow for the ventilation of the hose, floor of body removable over transmission, differential, etc. Front corners of body nicely curved interior of body strongly braced by means of steel braces, securely bolted to cross sills on bottom of body and riveted to panels. Capacity as desired.

Rear Step—Of hardwood, hung from rear end of body by hand forged braces. Step is 16" or deeper if preferred and as wide as the outside width of rear mudguards, covered with either linoleum, corrugated rubber matting, or aluminum, and bound on edges with metal binding. Directly in front of rear step and under body and chassis frame there is a hardwood tool box, with spring drop door and lock.

Railings—Heavy railings running from front end of body down to rear footboard and fastened there by means of sockets. Fastened to body by means of stanchions, front end of railings fitted with ornamental acorns. There is also a rail which serves as a hand hold for men riding on the rear step.

Driver's Seat—Of suitable size for two men, upholstered in genuine leather with spring back and cushions, stuffed with curled hair. Seat equipped with hand grabs on either side.

Chemical Tank—One 35 or 40-gallon "Pirsch Champion" or "Pirsch Standard" tank. The piping consists of three valves, a 2½" hose connection, chemical hose connection and a 2½" pressure gauge and overflow valve. The threads on the 2½" hose connection are cut to fit the city thread. All piping, tipping wheel, caps, etc., are of the best grade red brass, polished. All joints are properly sweated, threaded and soldered.

Chemical Tank Equipment—Two acid bottles, one acid bottle holder for running board, one soda bag and one wrench.

Chemical Charges—Two complete charges.

Ladders—One 12' Pirsch Pat. Trussed roof ladder with folding hooks attached; one 24' Pirsch Pat. Trussed extension ladder, rope hoist. One ladder on each side of body, or both on one side just as desired. Ladders finished in natural color of wood, ends black.

Hose Reel—One automatic chemical hose reel (or basket.)

Chemical Hose—150' ¾" chemical hose, fitted with heavy brass couplings, coupled in 50' lengths.

Chemical Nozzle—One chemical shut-off nozzle.

Extinguishers—Two 3-gallon Fire Dept. extinguishers in holders. Carried on running boards.

Lanterns—Two Dietz Fire Dept. lanterns in holders.

Bell—One 10" locomotive bell mounted on basket, dash or front of radiator or option of either one electric or one hand siren horn.

Searchlight—One 11" dash electric swivel searchlight.

Axe—One pick back fire axe in holders.

Crow Bar—One crow bar in holders.

Pike Pole—One 8' pike pole in holder.

Torches—Option of either two brass ornamental torches on rear standards or one Pirsch patent hose shut-off and door opener on side step.

Pump—The pump is of the rotary type with phosphor bronze rotars, alloy steel shafts, annular ball bearings, bronze stuffing box.

The drive unit consists of an oil-tight case carrying nickel steel hardened gears on annular ball bearings. The sliding gear is operated by a lever, which is on the outside of case convenient to the driver. This drive unit case is bolted to the pump head.

Each pump is equipped with two suction hose and two discharge hose connections, one of each on each side of car. The discharge connections are fitted with shut-off valves, while the suction connections are supplied with removable brass caps.

Relief Valves—One spring type or two Larkin.

Suction Hose—Two 10' lengths hard smooth bore suction hose with brass connections (size according to pump capacity).

Strainer—One brass suction hose strainer.

Gauges—One 300-pound pressure gauge; one combination vacuum and pressure gauge.

Reducer—One reducer.

Fenders—For rear wheels of sufficient size (unless furnished).

Side Steps and Driver's Footboards—To be covered with corrugated rubber matting, aluminum or linoleum with metal binding strip.

Painting and Lettering—Painting to be done in first-class manner, with best wearing coach colors and varnishes. Painted in any color desired; striping, lettering and decorations done in gold leaf with appropriate shadings. Lettering to order.

EVERYTHING FOR THE FIRE DEPT.

A Complete City Service Hook and Ladder Truck Equipment Mounted on Chassis

SPECIFICATIONS
"PIRSCH" City Service Hook and Ladder Truck Arches and Equipment

Seat—Special Fire Truck seat with genuine leather upholstering. Large enough to accomodate two men.

Running Boards—On each side as large as space will permit, from front fender to rear fender, and to rear fender.

Fenders—Rear fenders over rear wheels.

Tool Box—One on running board of ample size to carry all tools, etc.

Ladder Arches—One complete set of ladder arches with rubber covered rollers and flanges, equipped with automatic ladder lock. The arches are equipped with brass hand railing on both sides.

Ladders—One 50' Pirsch patent trussed extension ladder with rope hoist, automatic locks and stay poles.

One 40' Pirsch patent trussed extension ladder with rope hoist, automatic locks and stay poles.

One 28' Pirsch patent trussed single ladder.

One 24' Pirsch patent trussed single ladder.

One 20' Pirsch patent trussed single ladder.

One 16' Pirsch patent trussed single roof ladder.

One 12' Pirsch patent trussed single roof ladder.

One 16' Pirsch patent trussed inside extension ladder.

Two 25' Pirsch patent trussed single ladders on side.

Steel Basket—One perforated steel basket large enough to hold fireman's rubber coat, mounted on top of ladder arch.

Crow Bars—Two steel crow bars and holders.

Extinguishers—Two three-gallon Fire Department hand fire extinguishers with holders.

Fire Axes—Four heavy pick back Fire Department axes and holders.

Pike Poles—Two 12', two 14' and two 16' plaster hooks and pike poles and holders.

Crotch Poles—Two, with holders.

Wall Pick—One, with holder.

Shovels—Two, with holders.

Wire Cutter—One, with holder.

Pitch Forks—Two, with holders.

Tin Roof Cutter—One, with holder.

Chemical Tank—One or two, 40-gallon copper or steel chemical tanks, Pirsch Standard or Pirsch Champion type, with piping, gauge, valves, 2½" hose connection, 2 acid bottles, 1 brass canister for side step.

Hose Carrier—Either perforated steel hose basket with rollers or automatic hose reel.

Chemical Hose—200' ¾" 4-ply chemical hose.

Chemical Nozzle—One eccentric type shut-off nozzle.

Pull-Down Hook—One, with chain and hook.

Hose Shut-Off and Door Opener—One combination hose shut-off and door opener, "Pirsch" patent.

Life Line—One life line.

Lanterns—Four Fire Department hand lanterns and holders.

Locomotive Bell—Ten-inch locomotive bell.

Painting—English vermillion. All ladders natural wood with oil finish. All iron work black. Chassis painted English vermillion and gold leaf striped.

Lettering—As desired, in gold leaf.

Striping—In gold leaf.

The last chemical engine known to have been built was a five tank configuration on a 1934 Ford V-8 chassis for a Ridgeway Fire Department. Although the builder is not known, the rig has an American LaFrance bell, which hints at its possible origin.

VIII

THE MOST UNUSUAL CHEMICAL FIRE ENGINE

This illustration is obviously a water tower . . . . . but wait . . . . . could that be a chemical tank mounted in the middle of the chassis, and what is it doing there?

It is indeed a Babcock chemical tank, but, no, the chemicals do not go out of the nozzle at the top of the tower. The carbonic acid gas (carbon dioxide) generated when the acid is dumped into the soda water went into the two water-filled brass cylinders mounted horizontally on the chassis, and an incredible force of 10,000 to 20,000 pounds per square inch was transferred through hydraulics to pistons to *raise the water tower!*

Hale ⁂ Patent ⁂ Water ⁂ Tower.

It consists of a strong oak frame-work, mounted on wheels, carrying an iron frame with an extending telescopic tube, through which passes the hose, conducting the water from the supply to and through the pipe on the top end. The motor or lifting power is furnished by a chemical tank.

IN DETAIL.

THE FRAME is of well seasoned oak timber, four rails firmly fastened and bolted together, about 4 x 6 inches, 19 feet long, and re-inforced by plates of iron, ¼ x 5 inches. The frame, when put together, is about four feet wide. This constitutes the frame or body of the apparatus upon which rests the tower proper.

TANK, CYLINDERS, IRON SUPPORTS, ETC.

THE WHEELS are of the Archibald patent, of sufficient size consistent with strength, and capable of bearing the super-incumbent weight. The axles are of steel of the Concord pattern, and the whole is carried on platform springs.

THE TANK is the ordinary chemical one as used on chemical engines, of the Babcock or Halloway pattern, and develops a rising power of from 10,000 to 20,000 pounds. It is simple in its operation and action, and the power generated therein is conveyed at will into brass cylinders and thence to long piston rods attached to the quadrant.

THE QUADRANTS work in a movable rack of wrought iron attached to the piston rods, and, being fitted on the shaft of the tower, by their turning raise the tower into position.

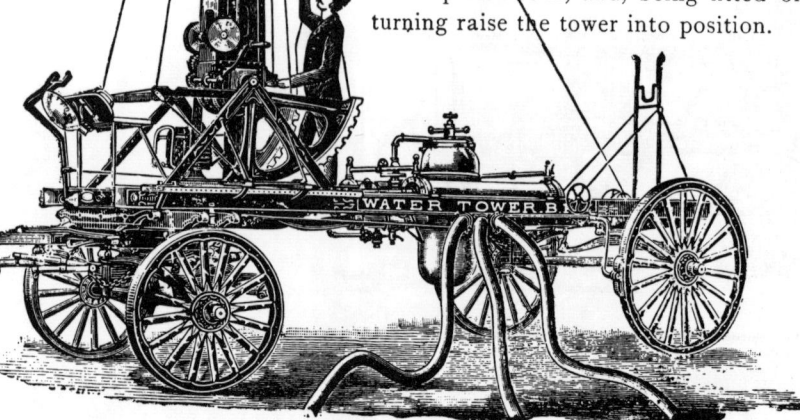

(Style 226.)
"HALE" PATENT WATER TOWER (without Deck-Turret.)

The Hale Patent Water Tower is the invention of George C. Hale, Chief of Fire Department of Kansas City, Mo., who is known to the fire-fighting civilized world as the inventor of many new and useful devices in the line of his profession.

The invention of the Water Tower, however, is the one that has brought his name most prominently before the public, the patriotic citizens of Kansas City, Mo., having made up a purse and sent him to the International Firemen's Meet, at London, some years since, Mr. Hale taking along with him one of the regular companies out of the Kansas City.

(Style 226.)

Fire Department, with their team of horses, Hale's Swinging Harness, a Hale Water Tower, etc., and his exhibition of quick hitching and getting out of the house were more than a seven days' wonder. On this trip Mr. Hale gave many exhibitions of the utility of his Water Tower.

The Hale Tower is elevated by the means of chemicals that are stored in a cylinder, which cylinder is nearly filled with water. When it is desired to place a tower in operation at a fire, the chemicals are mixed, pressure produced, and by opening a valve the pressure is exerted on two pistons, which are meshed into cogs of a segment. By this pressure the piston rods are moved, thereby elevating the Water Tower.

This tower is now in use in the following cities:

New York City, N. Y.,	Milwaukee, Wis.,	Omaha, Neb.,
Boston, Mass.,	Syracuse, N. Y.,	Minneapols, Minn.,
San Francisco, Cal.,	Louisville, Ky.,	Cincinnati, O.,
St. Joseph, Mo.,	Philadelphia, Pa.,	Baltimore, Md.,
Denver, Colo.,	New Orleans, La.,	St. Louis, Mo.,
Buffalo, N. Y.,	Kansas City, Mo.,	Lowell, Mass.

(Style 226.)

Is constructed to elevate to a height of 30 to 60 feet.
Will throw a stream horizontally a distance of 250 to 300 feet.

The nozzle is mounted on a turn-table which rests on top of base section of Water Tower, and is easily revolved by one man from the ground to any direction.

This tower is provided with a deck-turret or and-pipe r out five feet above the deck of the frame, and is arranged to throw a stream at any angle or point of the compass desired, cove: g cellar, first nd second floors of a building, having a sweep when throwing a stream of 250 feet in all directions. Is provided with 2, 2¼ and 2½-inch nozzles, and is easily m nipulated b one man.

IX

CHEMICALS LEAVE THE FIREHOUSE

It was inevitable, but Charles H. Fox of Ahrens-Fox did it first. He invented the booster system in 1913 that led to the demise of the chemical engine. The above photograph is *not* a chemical engine — it is Mr. Fox's first apparatus with a booster tank.

The new Ahrens-Fox booster system had a small centrifugal pump mounted in front of the radiator, and was connected to a water tank behind the driver's seat (today this tank is still known as the "booster tank") which replaced the chemical tank. Ahrens-Fox delivered the first ten fire engines with booster systems during 1913. All ten of them went to the Cincinnati Fire Department, Mr. Fox's former employer.

Always an innovator, Charles H. Fox left the Cincinnati Fire Department in 1908 to join John P. Ahrens in a new fire apparatus manufacturing firm to be known as Ahrens-Fox Fire Engine Co. Their bread and butter item was the Continental steamer. In 1910 they brought out the battery-powered Continental fire chief's auto with chemical tank (see Chapter VII) and in 1911 their first motorized engine with a front mounted piston pump.

Then came steamers propelled by either electrical batteries or gasoline engines, but the end of the steamer era was at hand. Although the three hybrids (battery/chemical, battery/steam, gasoline engine/steam) never were accepted to any great extent, it was the combination of an internal combustion engine to propel the apparatus as well as to power both a volume pump and a smaller booster pump which led to the "triple combination pumper", the backbone of America's fire service from the 1920's until the present day. Thus Charles Fox was first to hit upon the right combination which has endured.

Yet the transition from chemical to booster systems was not overnight. It took place over two decades, with the last known chemical engine being manufactured in 1934. As late as 1927 the Ahrens-Fox Company themselves continued to provide, on their own apparatus, chemical tanks manufactured by Peter Pirsch & Sons.

It was far easier, simpler, and less expensive to use a booster pump on a water tank than to charge a tank of soda water with messy and dangerous sulphuric acid. But this simplicity does not address the fact that for half a century the "chemicals" were believed to be 30 to 40 times more effective than plain water. The sad truth was, of course, that the chemicals were in fact no more effective than plain water. The chemicals, other than providing the force to expel the water, merely acted as a psychological placebo for the firefighters.

An article in the National Fire Protection Association Fire Almanac entitled "Chemical Fire Apparatus" states, "So at first it was thought that all hand and steam fire engines could be replaced with chemical engines carrying 100 to 200 gallons of water. When the water was mixed with 'magic' chemicals and projected through a one inch hose, these engines could extinguish any fire using a 2½ inch hose stream. But chemical engines were overrated and soon their limitations were realized."

The Round Table Department in the June 1932 issue of *Fire Engineering* considered the chemical tank versus booster tank controversy. The vote of 28 chiefs was 27 for booster, 1 for chemical. The chiefs responded to the following three questions:

1. From experience in fighting fires, which would you prefer: a 35- or 40-gallon chemical tank on your apparatus, or a booster tank?

2. What do you consider the advantage and disadvantage of each?

3. Do you find that chemical lines do better work than small lines from the booster tank, or vice versa?

Their reasons for favoring the booster system may be summarized as follows:

1. The acid and the sodium bicarbonate solutions would not mix uniformly and instantly. Therefore, the initial stream would contain acid which would corrode the hose fittings, rot the hose, damage the firemen's clothing, and make holes in the objects to which the stream was applied, such as fabrics on furniture, beds, etc. The word "messy" was used repeatedly in the chiefs' responses. In comparison, the booster tank was a relatively simple, clean operation.

2. If the chemical tank was activated, the entire tank (generally 35 or 40 gallons) had to be discharged even though the fire might be extinguished with 10 gallons. Any amount could be used from the booster tank.

3. If more than one tank was required, recharging on the scene was difficult and slow. The tank had to be refilled with water, the soda added, and then the bottle of acid had to be put in place in the tank. Any acid spilled on the rig dissolved paint. If the chemical tank were over the hose bed, any spilled acid quickly destroyed the cotton hose fabric. A booster tank could be refilled with garden hoses, bucket brigades, or supply lines, if available.

4. There was no control on the pressure of the chemical stream. Once the reaction was initiated, the pressure buildup was strictly a function of the mixing rate of the reacting chemicals. If the nozzle were to be shut down, the pressure would increase and possibly burst the hose. The booster tank pressure could be controlled by the pump operator.

5. The cost of the water for filling the booster tank was nil, while there was a direct cost for the chemicals used in the chemical tank.

6. Several chiefs mentioned that they felt the chemical lines were more effective in small enclosed areas. (Even in 1932 the myth still persisted to a degree).

In his article entitled, "Copper & Brass — The Chemical Engine Era!" in the August 1959 issue of *Fire Engineering*, Clarence E. Meeks, who was himself assigned to a horse drawn chemical engine early in his career, states, "But like most all new and revolutionary ideas, the chemicals were vastly overrated, so much so, that many fire departments had to fight against a wide-spread belief that all hand and steam fire engines could be replaced with chemical engines."

In regard to the new "booster" apparatus invented by Mr. Fox, Clarence Meeks states in his *Fire Engineering* article, "While the idea of putting a small stream of plain water on a fire through a "garden hose" was openly ridiculed by many fire officers who still thought that 'chemicals' had extinguishing qualities not possessed by ordinary water, the 'booster' grew rapidly in favor and soon displaced the chemical engines and chemical combinations."

Mr. Meeks attended the 1913 convention of the International Association of Fire Engineers at the Grand Central Palace in New York City at which the first Ahrens-Fox "booster" engine was introduced. Meeks, then a New York City Battalion Chief, overheard another chief exclaim, "That thing will never replace the chemical engine." But within two decades the new booster system had indeed replaced chemical apparatus, and the curtain had come down on the chemical engine era.

Although the myth of the chemicals was finally dispelled, the worth of chemical engines can never be underestimated. In spite of their misleading and false claims, they *did* save countless millions of dollars worth of property, and no doubt many lives would have been lost without them.

Chemical fire engines — misunderstood, overrated, but none the less one of the heroes — perhaps the most unsung hero — of our romantic heritage from the American fire service.

WITHDRAWN

No longer the property of the
Boston Public Library.
Sale of this material benefited the Library.

Boston Public Library

COPLEY SQUARE
GENERAL LIBRARY

TH9375
.C65
1987
87033310-01

The Date Due Card in the pocket indicates the date on or before which this book should be returned to the Library.

Please do not remove cards from this pocket.